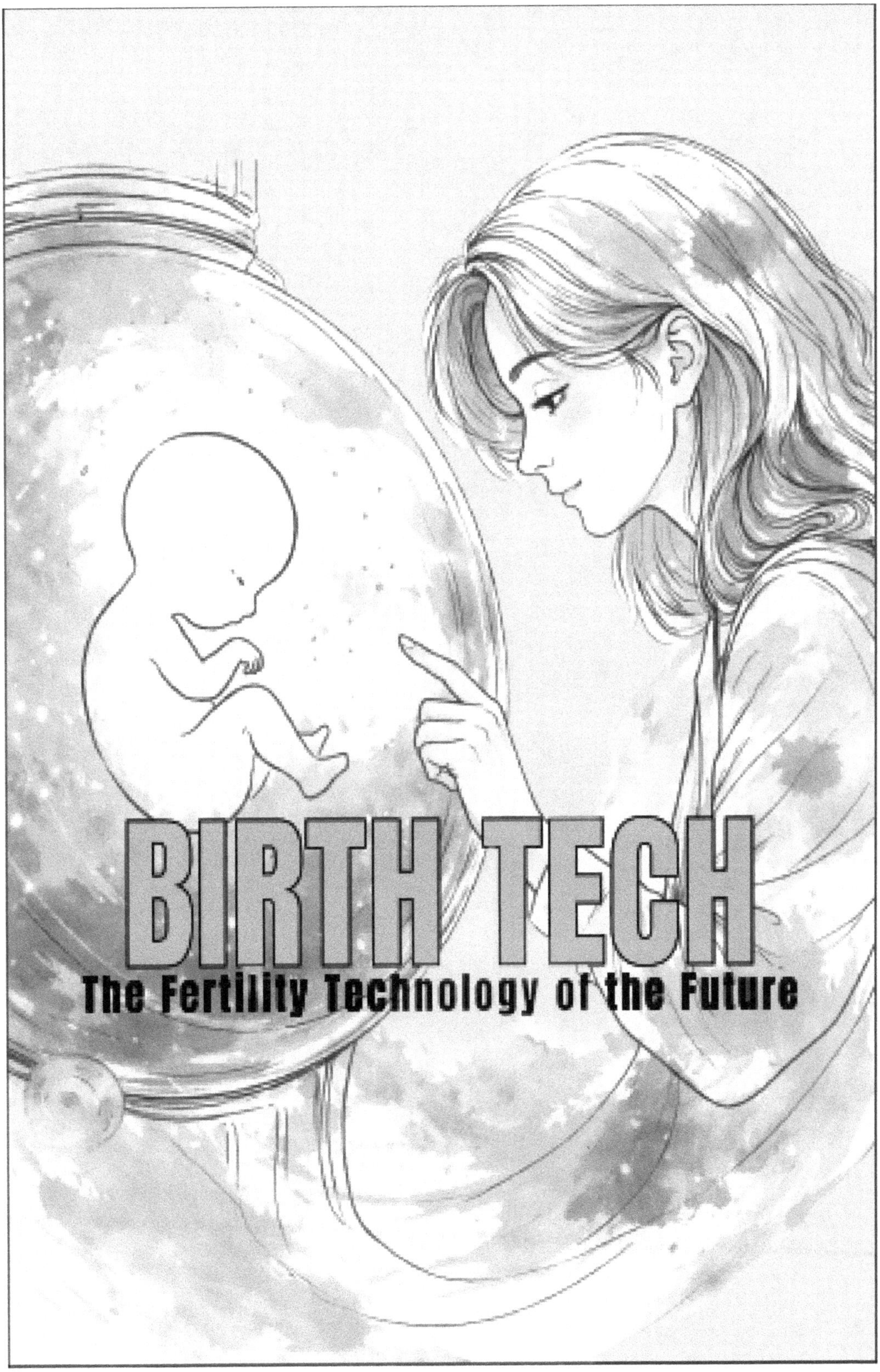
BIRTH TECH
The Fertility Technology of the Future

BIRTH TECH

The Fertility Technology Of The Future

By

Akasha Rehn & PRSC Whitley

For TechEyeSpy

COPYRIGHT

THIS BOOK IS NOT MEDICAL ADVICE. IT REFLECTS REPORTING, RESEARCH, AND LIVED EXPERIENCE AT THE TIME OF WRITING. MEDICAL DECISIONS SHOULD
ALWAYS BE MADE WITH QUALIFIED PROFESSIONALS.

THIS BOOK IS NOT FINANCIAL ADVICE. IT REFLECTS REPORTING, RESEARCH, AND LIVED EXPERIENCE AT THE TIME OF WRITING.FINANCIAL DECISIONS SHOULD
ALWAYS BE MADE WITH QUALIFIED PROFESSIONALS.

PAPERBACK ISBN: 978-1-9193517-0-4

10 9 8 7 6 5 4 3 2 1

Dedication

For those who assumed time would wait.
And for those who discovered it doesn't, quietly.

First Edition Notice

This work is published as a first print edition.

The text has been written and structured for long-form reading in physical format. Layout, typography, and pagination have been adapted specifically for print, and references and appendices are presented to support linear reading.

This edition reflects the state of research, regulation, and public discourse at the time of writing. As with all works engaging with active scientific and technological fields, future editions may revise or expand sections to reflect new evidence, clinical outcomes, or regulatory developments.

AUTHOR'S NOTE

THIS IS NOT A MEMOIR.

IT IS NOT A GUIDE, A MANIFESTO, OR AN ARGUMENT. IT DOES NOT ATTEMPT TO SPEAK FOR WOMEN, FOR FAMILIES, OR FOR ANYONE BEYOND THE PERSON OBSERVING WHAT UNFOLDS ON THESE PAGES.

THIS BOOK FOLLOWS TWO PARALLEL THREADS.

ONE IS AN EXPLORATION OF FERTILITY TECHNOLOGIES AS THEY ACTUALLY EXIST: FROM THE ORDINARY SYSTEMS MOST PEOPLE NOW TAKE FOR GRANTED TO THE SPECULATIVE RESEARCH QUIETLY RESHAPING WHAT REPRODUCTION MAY BECOME. THESE CHAPTERS ARE WRITTEN TO EXPLAIN WHAT THESE TECHNOLOGIES DO, WHY THEY EXIST, AND WHAT CHANGES WHEN THEY WORK.

THE OTHER THREAD IS MORE INCIDENTAL. IT FOLLOWS ONE PERSON MOVING THROUGH THIS LANDSCAPE WITHOUT EXPECTING TO STAY. THE PERSONAL MOMENTS THAT APPEAR ARE NOT OFFERED AS REPRESENTATIVE, EXEMPLARY, OR INSTRUCTIVE. THEY EXIST ONLY TO ANCHOR ABSTRACTION TO REALITY.

NOTHING HERE ARGUES THAT PEOPLE SHOULD HAVE CHILDREN EARLIER, LATER, OR AT ALL. NOTHING HERE SUGGESTS THAT BIOLOGY IS DESTINY OR THAT TECHNOLOGY IS SALVATION. IT SIMPLY RECOGNISES THAT PROBABILITY OPERATES WHETHER WE ACKNOWLEDGE IT OR NOT, AND THAT AWARENESS OFTEN ARRIVES AFTER IT WOULD HAVE BEEN EMOTIONALLY NEUTRAL.

WHERE HUMOUR APPEARS, IT DOES SO QUIETLY. WHERE DISCOMFORT APPEARS, IT IS ALLOWED TO REMAIN UNRESOLVED.

THE READER IS TRUSTED TO DRAW THEIR OWN CONCLUSIONS.

Preface

On Health, Time, and Odds

There was nothing wrong with my health. My fertility, on paper, was fine.

That sentence is easy to misunderstand. It sounds reassuring, and for a long time it was. In my twenties and early thirties, it functioned as a kind of background assumption, the way good weather does when you don't check the forecast. Healthy meant safe. Fine meant flexible. Time felt neutral.

Only later did I realise that health and fertility are not guarantees so much as conditions under which probability still applies. The odds were never hostile, but they were never static either. They had been shifting quietly for years, unnoticed because nothing had gone wrong yet.

I did not regret the life I lived. I did not wish it undone. What unsettled me was recognising how little emphasis had been placed on understanding the mathematics while it was cheap to understand them. By the time the numbers began to matter, they had already been moving for some time.

This book begins there, not because regret is instructive, but because clarity often arrives late. What follows is not an attempt to recover lost time, but to examine what fills the space once illusion falls away.

The technologies discussed in these pages grow progressively more complex, more abstract, and less anchored to the human body as we currently understand it. That escalation mirrors the journey most people take unintentionally: from ordinary assumptions, through uncommon interventions, and finally into ideas that feel distant enough to be optional until they are not.

At a certain point, the research stops being personal. Curiosity replaces urgency. Observation replaces need. That shift matters, and it shapes how this story unfolds.

This is where we start.

Chapters

Chapter One

On Time, Probability, and Realising Too Late

Before the System Noticed Me

I've always paid attention to the worlds systems, even when I didn't think of myself as technical.

Not in the heroic sense. I wasn't taking machines apart or teaching myself to code at night. I just noticed patterns. Timetables. Interfaces. The small frictions that told you how something really worked, not how it was described.

Dating, for example, never felt romantic to me so much as procedural. Even before apps, there were rules. Who approached whom. How long you waited. What counted as interest and what counted as rudeness. By the time the internet formalised all of this, it felt less like a revolution and more like an admission. The system had been there all along. We'd just stopped pretending otherwise.

The first time I used an online dating platform, what struck me wasn't the people. It was the filters. Age ranges. Distance sliders. Education checkboxes. A quiet confidence that these variables mattered, that they could be adjusted, optimised, improved. No one mentioned fertility, of course. It hovered in the background, implied by numbers rather than words.

At the time, I didn't think of this as reproductive technology. It didn't touch the body. It didn't require consent forms or blood tests. It just rearranged probability. Who met whom. When. Under what assumptions.

That felt benign. Sensible, even.

I liked the clarity. I liked that preferences were declared rather than guessed. I liked that rejection arrived cleanly, without explanation. The interface didn't care how you felt about being overlooked. It just moved on. There was something reassuring about that efficiency.

Looking back, it's obvious this was my first real encounter with fertility tech. I just didn't have the language for it yet.

At this stage, nothing felt urgent. Health was fine. Life was busy. Time felt like a resource you could dip into later. The idea that the odds were already in motion didn't register because nothing had yet asked me to notice.

What I did notice, though, were the quiet assumptions built into the platforms. The way age was treated as neutral until it wasn't. The way seriousness was signalled. The way intentions were framed as lifestyle choices rather than biological ones. Nobody lied. They just didn't say certain things out loud.

That suited me.

There's a comfort in systems that don't editorialise. They don't tell you what matters. They let you discover it slowly, sometimes too slowly, by watching what changes when you adjust a variable.

At the time, I assumed this was about convenience. Efficiency. Modernity. I didn't see it as the first layer in a much larger stack, one that would eventually extend from algorithms into laboratories and then beyond the human body altogether.

But the habit was already there. The habit of trusting systems to surface what mattered. The habit of believing that if something was important, it would eventually be made explicit.

That belief would not survive the book you're about to read.

At this point, though, none of that felt dramatic. It felt normal. Harmless. Even dull. Fertility wasn't a concern. Technology was just infrastructure. And probability was something that applied to other people, later.

That was the illusion. And it worked perfectly, right up until it didn't.

Was time on my side?

Chapter Two

Matching Before Meaning

Internet Dating Platforms as Reproductive Infrastructure

2.1 Internet Dating Platforms as Probabilistic Reproductive Infrastructure

Online dating platforms are not social curiosities. They are large-scale probabilistic systems that mediate mate selection across entire populations. Their relevance to fertility does not arise from intent, but from effect. By controlling who encounters whom, when, and under what filtering constraints, they alter the timing, composition, and durability of reproductive pairings long before biology or medicine is involved.

By the late 2010s, internet dating had become the dominant pathway through which couples formed in most high-income countries. A landmark demographic study by Michael Rosenfeld and colleagues at Stanford University, published in the Proceedings of the National Academy of Sciences in 2019, demonstrated that online platforms had overtaken family, friends, education, and workplace introductions as the primary way heterosexual couples met in the United States. The same trend has been observed across the UK, Western Europe, Canada, Australia, and parts of East Asia, with minor cultural variation.

This shift matters because reproduction is downstream of pairing. Any system that systematically delays, accelerates, fragments, or stratifies pair formation will inevitably shape fertility outcomes, regardless of whether fertility is named as a design goal.

The online dating ecosystem is not pluralistic in practice. It is highly concentrated.

At the centre of this concentration sits Match Group, a publicly listed company that, through acquisition and platform differentiation, controls a substantial portion of global online dating traffic. Match Group's portfolio includes Tinder, Hinge, OkCupid, Plenty of Fish, Match.com, and several regional services. According to Match Group investor disclosures and annual reports between 2019 and 2023, the company has consistently accounted for over half of global online dating revenue, with Tinder alone representing the majority of active users.

This level of consolidation means that a small number of design philosophies now mediate a large share of reproductive introductions.

The most influential of these philosophies is high-velocity matching.

Tinder, launched in 2012, introduced the swipe interface that has since become industry standard. From a technical perspective, this interface optimises for throughput. Users are encouraged to

evaluate profiles rapidly, often in seconds, based primarily on visual cues and minimal text. Behavioural research published in Computers in Human Behavior (Ward, 2016) found that Tinder users commonly assessed dozens to hundreds of profiles per session, a scale of evaluation impossible in offline contexts.

High-throughput evaluation has predictable systemic consequences. It increases choice density while reducing commitment friction. It raises standards while lowering switching costs. Sociological and economic studies of online matching markets, including work by Hitsch, Hortaçsu, and Ariely published in the American Economic Review (2010), show that when choice sets expand dramatically, users become more selective, not less, and are more likely to reject viable matches in anticipation of marginally better ones.

From a fertility perspective, this selectivity interacts with time asymmetrically.

Age is one of the most commonly used filters across dating platforms. While framed as preference, age filtering functions as a hard constraint on reproductive timelines. Most platforms default to narrow age bands, often five years either side of the user's stated age. These defaults are rarely questioned, yet they significantly shape who is visible to whom at different life stages.

The effect is cumulative. Years spent inside algorithmically bounded dating pools are years in which pairing probabilities are altered without any explicit
decision to delay reproduction.

Importantly, platforms do not monetise successful exits. They monetise continued engagement.

Match Group's revenue model is driven primarily by subscriptions, premium features, and in-app purchases. Public filings show that revenue growth correlates strongly with user retention and time spent on platform, not with the rate at which users form long-term relationships. This creates an inherent tension between platform incentives and reproductive efficiency. While no platform actively discourages partnership, none structurally prioritise early exit as a success metric.

This is not a moral claim. It is an incentive alignment problem.

THE RESULT IS A SYSTEM THAT IS HIGHLY EFFECTIVE AT FACILITATING INTRODUCTIONS WHILE BEING STRUCTURALLY INDIFFERENT TO BIOLOGICAL TIME. THE PLATFORMS DO NOT NEED TO ENCOURAGE DELAY; THEY SIMPLY DO NOT PRICE IT.

BY THE LATE 2010S, THIS MODEL BEGAN TO GENERATE VISIBLE SECOND-ORDER EFFECTS. RISING AGE AT FIRST LONG-TERM PARTNERSHIP, SHORTER AVERAGE RELATIONSHIP DURATION PRIOR TO COHABITATION, AND PROLONGED PERIODS OF SERIAL DATING HAVE BEEN DOCUMENTED IN SOCIOLOGICAL LITERATURE ACROSS MULTIPLE COUNTRIES. THESE TRENDS INTERSECT DIRECTLY WITH FEMALE FERTILITY DECLINE, WHICH ACCELERATES STATISTICALLY AFTER THE EARLY THIRTIES DESPITE INDIVIDUAL VARIATION IN HEALTH.

At no point does the system flag this interaction explicitly.

What makes online dating particularly consequential as fertility infrastructure is that it operates before awareness. Users engage with these platforms under the assumption that fertility is a distant or future concern. The system benefits from this assumption. By the time fertility becomes salient, years of probabilistic filtering have already taken place.

The platforms themselves are not static. In response to criticism of swipe culture and declining user satisfaction, Match Group introduced differentiated products such as Hinge, which slows interaction through prompts, reduced match volume, and explicit signalling of relationship intent. Hinge markets itself as "designed to be deleted," an acknowledgement that unlimited choice can undermine durability.

However, even this corrective move operates within the same probabilistic framework. It refines selection rather than escaping it. It remains an optimisation problem, not a biological one.

What emerges, taken as a whole, is a pre-clinical reproductive system that shapes fertility outcomes without appearing to do so. It selects, delays, stratifies, and redistributes pair formation across time and population groups, all while presenting itself as lifestyle software.

This is why internet dating belongs at the start of any serious discussion of fertility technology. Not because it is controversial, but because it is foundational. It determines who reaches the clinic, when they arrive, and with whom. By the time medical intervention is considered, the largest probabilistic shifts have already occurred upstream, inside systems most users never think to question.

References

Rosenfeld, M. J., Thomas, R. J., & Hausen, S. (2019). Disintermediating your friends: How online dating in the United States displaces other ways of meeting. Proceedings of the National Academy of Sciences.

Hitsch, G. J., Hortaçsu, A., & Ariely, D. (2010). Matching and sorting in online dating. American Economic Review.

Ward, J. (2016). What are you doing on Tinder? Impression management on a matchmaking mobile app. Computers in Human Behavior.

Match Group Inc. Annual Reports and Form 10-K filings, 2019–2023.

2.2 Demographic Pressure, Pairing Collapse, and the Statistical Case for Algorithmic Matching

Internet dating platforms did not arise in a demographic vacuum. They are not an arbitrary technological intrusion into social life. They emerged in response to measurable, compounding demographic shifts that made traditional pairing mechanisms insufficient at scale. The technology did not create the problem it addresses. It capitalised on it.

At the most basic level, human reproduction requires pairing within a narrowing temporal window. Any society that disrupts the conditions under which pairing reliably occurs will eventually face fertility consequences, regardless of cultural intent. By the late twentieth century, this disruption was already well underway.

Across high-income countries, three trends converged simultaneously.

First, age at first partnership and marriage rose sharply. In the United Kingdom, the Office for National Statistics reports that the average age at first marriage for women increased from approximately 23 years in 1971 to over 31 years by 2021, with men averaging closer to 33. Similar patterns are observed across Western Europe, North America, Australia, and parts of East Asia. This shift did

NOT OCCUR BECAUSE PEOPLE STOPPED FORMING RELATIONSHIPS, BUT BECAUSE THE TRANSITION FROM CASUAL ASSOCIATION TO DURABLE PAIRING SLOWED.

SECOND, THE INSTITUTIONAL ENVIRONMENTS THAT PREVIOUSLY FACILITATED PAIRING WEAKENED. WORKPLACES BECAME MORE TRANSIENT, EDUCATION LASTED LONGER, AND GEOGRAPHIC MOBILITY INCREASED. THE PROPORTION OF ADULTS MEETING PARTNERS THROUGH FAMILY, COMMUNITY, OR RELIGIOUS INSTITUTIONS DECLINED STEADILY THROUGHOUT THE LATE TWENTIETH CENTURY. MICHAEL ROSENFELD'S DEMOGRAPHIC WORK, DRAWING ON U.S. LONGITUDINAL SURVEY DATA, SHOWS A MARKED COLLAPSE IN INTRODUCTIONS MEDIATED BY FRIENDS AND FAMILY BETWEEN THE 1980S AND 2010S, REPLACED PRIMARILY BY ONLINE SYSTEMS.

THIRD, SEX RATIOS WITHIN KEY AGE COHORTS BECAME UNEVEN, PARTICULARLY IN URBAN CENTRES. LARGE CITIES WITH STRONG SERVICE, FINANCE, AND TECHNOLOGY SECTORS OFTEN EXHIBIT FEMALE-SKEWED POPULATIONS IN EARLY ADULTHOOD AND MALE-SKEWED POPULATIONS AT LATER CAREER STAGES, DEPENDING ON MIGRATION PATTERNS. THESE IMBALANCES REDUCE THE EFFICIENCY OF LOCAL PARTNER SEARCH AND INCREASE COMPETITION WITHIN NARROW DEMOGRAPHIC BANDS.

TAKEN TOGETHER, THESE FORCES CREATED A PAIRING PROBLEM.

THE PROBLEM WAS NOT MORAL, IDEOLOGICAL, OR POLITICAL. IT WAS MATHEMATICAL.

If individuals are dispersed, mobile, time-poor, and delaying commitment while still desiring long-term partnership eventually, then informal matching systems fail. The probability of encountering a compatible partner through random social exposure drops below a viable threshold. At that point, structured mediation becomes rational.

This is the demographic niche that internet dating occupies.

Online platforms scale partner search across geography, time, and social networks. They compensate for the loss of dense, overlapping communities by replacing them with algorithmic proximity. In doing so, they convert a collapsing social function into a service.

The numbers justify this intervention.

By the late 2010s, surveys across the UK, United States, and EU indicated that between 35 and 45 percent of adults under 40 had used an online dating platform, with usage rates exceeding 60 percent in major urban centres. Among never-married adults, usage was higher still. These figures are not marginal. They describe a dominant pairing pathway.

From a fertility perspective, the relevance becomes clearer when age-specific birth rates are considered. In the UK, the Office for National Statistics reports that fertility rates among women aged 30–34 overtook those aged 25–29 for the first time in 2013, while rates among women under 25 declined sharply. This shift reflects delayed partnership formation rather than declining desire for children. Surveys consistently show that ideal family size has remained relatively stable, typically between 1.8 and 2.2 children, depending on country.

The gap, therefore, is not aspiration. It is execution under compressed time.

Internet dating responds to this compression by increasing encounter frequency, not biological efficiency. It allows individuals to meet more potential partners later in life, but it cannot reverse the underlying decline in fecundity associated with age. What it can do is reduce search friction when time is already constrained.

This is where algorithmic filtering becomes central.

Platforms allow users to specify age ranges, education levels, relationship intentions, and lifestyle preferences. From an individual perspective, this feels empowering. From a population perspective, it produces assortative mating at scale.

Assortative mating, the tendency for individuals to pair with others who share similar traits, is not new. What is new is its precision and enforcement. Research in demography and sociology shows increasing educational homogamy in developed countries, with graduates increasingly partnering with other graduates. This trend correlates strongly with delayed childbearing, as extended education postpones stable partnership formation and compresses reproductive timelines.

Online platforms amplify this effect by making educational status an explicit filter rather than an emergent pattern. The same applies to income proxies, cultural background, and declared family intentions. The system does not invent preferences. It hardens them into gates.

From a fertility systems perspective, this produces two outcomes simultaneously.

For some users, pairing becomes more efficient. Compatible partners who might never have met offline are introduced quickly, potentially enabling family formation that would otherwise not occur.

For others, the narrowing of acceptable matches delays pairing further, especially as age increases and filters become mutually restrictive. The pool shrinks, not because partners do not exist, but because algorithmic constraints intersect.

This dual effect explains why internet dating is both ubiquitous and frequently dissatisfying.

Demographically, however, the system is justified by the numbers. Without algorithmic mediation, pairing rates among mobile, urban, late-marrying populations would likely fall further, exacerbating already declining fertility rates. Governments implicitly recognise this, even if they do not state it plainly. No major high-income country has attempted to restrict online dating. Some have actively partnered with platforms for public health messaging, recognising their reach.

The technology persists because it solves a real coordination problem, even if imperfectly.

What it does not solve is biological asymmetry. Time compression affects men and women differently. While male fertility declines more gradually, female fertility declines more steeply after the early thirties. Internet dating platforms treat time symmetrically. Age is a filter, not a biological warning system. The system assumes rational actors with perfect information. Real users rarely possess either.

This is not a failure of ethics. It is a limitation of abstraction.

By the time fertility becomes a conscious concern, the demographic logic that justified internet dating has already done its work. The platforms have enabled pairing later in life, but they have also normalised delay as manageable. When that assumption collides with biological reality, the next layer of fertility technology enters the picture.

Clinics do not replace dating platforms. They inherit their outcomes.

References

Office for National Statistics (UK). Births in England and Wales statistical series, 2010–2023.

Rosenfeld, M. J., Thomas, R. J., & Hausen, S. (2019). Disintermediating your friends: How online dating in the United States displaces other ways of meeting. Proceedings of the National Academy of Sciences.

Eurostat. Mean age at first marriage and fertility indicators, EU-27 datasets.

United Nations Department of Economic and Social Affairs. World Fertility Data summaries.

American Sociological Review and demographic literature on assortative mating and educational homogamy, 2000–2020.

2.3 Monetisation, Engagement Loops, and Time Asymmetry in Digital Pairing Systems

Internet dating platforms do not merely mediate introductions. They are monetised engagement systems operating under asymmetric temporal constraints. This asymmetry is not ideological or intentional; it is structural. The platforms are designed to optimise user activity over time, while fertility outcomes are constrained by time in a non-linear and biologically uneven way.

This tension sits at the core of modern fertility infrastructure.

Digital dating platforms operate on a business model familiar from other consumer technology sectors: maximise retention, increase interaction frequency, and convert engagement into recurring revenue. Subscription tiers, premium visibility boosts, read receipts, algorithmic prioritisation, and profile enhancement features are all mechanisms designed to keep users active within the system.

Public financial disclosures from Match Group make this explicit. Across its portfolio, revenue growth is driven primarily by paid subscriptions and in-app purchases rather than advertising. Annual reports between 2019 and 2023 repeatedly emphasise metrics such as monthly active users, payer penetration, and average revenue per user.

None emphasise successful relationship formation as a performance indicator.

This is not evidence of malice. It is evidence of incentive alignment.

A platform that reliably paired users quickly and permanently would, by definition, reduce its own addressable market. The business logic therefore favours systems that facilitate ongoing participation rather than rapid exit. This does not require deliberate obstruction of relationships. It merely requires neutrality toward outcomes once engagement is secured.

The resulting environment encourages what behavioural economists describe as continuous optionality. Users remain open to alternatives even after forming promising connections. The marginal cost of disengaging from a current interaction in favour of a hypothetical better one is low. This dynamic is amplified by interface design choices that foreground abundance rather than scarcity.

From a fertility perspective, the consequences of this design are not evenly distributed across time or sex.

Biological fertility does not decline linearly. In women, fecundity remains relatively stable through the twenties, begins a gradual decline in the early thirties, and accelerates thereafter. This pattern is well documented in

EPIDEMIOLOGICAL AND DEMOGRAPHIC LITERATURE AND REFLECTED IN NATIONAL BIRTH STATISTICS ACROSS DEVELOPED COUNTRIES. MALE FERTILITY DECLINES MORE GRADUALLY AND REMAINS VIABLE ACROSS A LONGER AGE RANGE, THOUGH NOT INDEFINITELY.

DIGITAL DATING PLATFORMS, BY CONTRAST, TREAT TIME AS FUNGIBLE. A YEAR SPENT BROWSING PROFILES AT TWENTY-SEVEN IS FUNCTIONALLY EQUIVALENT TO A YEAR SPENT BROWSING AT THIRTY-SEVEN, FROM THE SYSTEM'S PERSPECTIVE. ENGAGEMENT METRICS DO NOT DIFFERENTIATE BETWEEN THESE STATES. THE INTERFACE DOES NOT ESCALATE URGENCY. THE SYSTEM DOES NOT SURFACE PROBABILISTIC WARNINGS. TIME IS INVISIBLE.

THIS INVISIBILITY IS CENTRAL TO THE PLATFORM'S APPEAL.

USERS ARE NOT ASKED TO CONFRONT BIOLOGICAL TRADE-OFFS WHILE INTERACTING WITH DATING APPS. THE EXPERIENCE IS FRAMED AS LIFESTYLE MANAGEMENT RATHER THAN LIFE PLANNING. PROFILES EMPHASISE INTERESTS, VALUES, AND AESTHETICS. FERTILITY IS RELEGATED TO OPTIONAL FIELDS, OFTEN BINARY AND POORLY SPECIFIED, SUCH AS "WANTS CHILDREN" OR "OPEN TO CHILDREN," WITHOUT REFERENCE TO TIMING OR PROBABILITY.

THE PLATFORM THUS MAINTAINS ENGAGEMENT BY DECOUPLING DESIRE FROM CONSTRAINT.

EMPIRICAL RESEARCH SUPPORTS THE CLAIM THAT THIS DECOUPLING PROLONGS PARTNER SEARCH. STUDIES PUBLISHED IN THE JOURNAL OF SOCIAL AND PERSONAL RELATIONSHIPS AND SOCIOLOGICAL SCIENCE DURING

THE 2010S INDICATE THAT ONLINE DATERS, ON AVERAGE, SPEND LONGER PERIODS IN NON-COMMITTAL INTERACTION PHASES THAN THOSE MEETING THROUGH OFFLINE NETWORKS, EVEN WHEN CONTROLLING FOR AGE AND EDUCATION. THE EFFECT IS MOST PRONOUNCED IN URBAN POPULATIONS WITH HIGH PLATFORM PENETRATION.

THIS PROLONGED SEARCH PHASE INTERACTS DIRECTLY WITH FERTILITY TIMELINES.

BY THE TIME USERS BEGIN TO EXPERIENCE TIME PRESSURE EXPLICITLY, THE HABITS OF ENGAGEMENT ARE ALREADY ESTABLISHED. CONTINUOUS BROWSING, ITERATIVE OPTIMISATION OF PROFILES, AND INCREMENTAL REFINEMENT OF PREFERENCES FEEL NORMAL. THE SYSTEM HAS TRAINED USERS TO BELIEVE THAT BETTER MATCHES REMAIN AVAILABLE WITH SUFFICIENT PATIENCE.

FROM A MARKET PERSPECTIVE, THIS BELIEF SUSTAINS DEMAND.

FROM A FERTILITY PERSPECTIVE, IT CREATES A MISMATCH BETWEEN PERCEIVED AND ACTUAL TEMPORAL FLEXIBILITY.

IT IS IMPORTANT TO STRESS THAT PLATFORMS DO NOT INVENT DELAYED REPRODUCTION. DELAYED CHILDBEARING PREDATES WIDESPREAD INTERNET DATING AND IS DRIVEN BY EDUCATION, LABOUR MARKET PARTICIPATION, HOUSING COSTS, AND CULTURAL CHANGE. HOWEVER, DATING PLATFORMS NORMALISE DELAY AS MANAGEABLE BY PROVIDING A CONSTANT STREAM OF POTENTIAL CONNECTIONS REGARDLESS OF AGE.

THIS NORMALISATION IS REINFORCED BY MONETISATION STRATEGIES THAT REWARD PERSISTENCE.

PREMIUM TIERS OFTEN PROMISE INCREASED VISIBILITY, HIGHER MATCH RATES, OR ALGORITHMIC PRIORITISATION. THESE FEATURES IMPLICITLY SUGGEST THAT INSUFFICIENT SUCCESS IS A PROBLEM OF EXPOSURE RATHER THAN TIMING. THE SOLUTION OFFERED IS NOT URGENCY, BUT UPGRADE.

IN ECONOMIC TERMS, THIS CONVERTS TIME PRESSURE INTO CONSUMER FRICTION.

FROM A FERTILITY SYSTEMS PERSPECTIVE, THIS IS SIGNIFICANT BECAUSE IT DISPLACES RESPONSIBILITY. RATHER THAN CONFRONTING DECLINING ODDS, USERS ARE ENCOURAGED TO OPTIMISE WITHIN THE SYSTEM. BETTER PHOTOS. BETTER PROMPTS. BETTER FILTERS. BETTER SUBSCRIPTIONS. THE UNDERLYING BIOLOGICAL CLOCK REMAINS UNADDRESSED.

THIS DISPLACEMENT DOES NOT AFFECT ALL USERS EQUALLY.

WOMEN EXPERIENCE A SHARPER DIVERGENCE BETWEEN SYSTEM-ENCOURAGED PATIENCE AND BIOLOGICAL REALITY. MEN, ON AVERAGE, FACE A LONGER VIABLE REPRODUCTIVE WINDOW AND THUS EXPERIENCE LESS IMMEDIATE COST FROM EXTENDED SEARCH. THE PLATFORM TREATS BOTH SYMMETRICALLY, BUT BIOLOGY DOES NOT.

THE RESULT IS NOT EXPLOITATION IN A LEGAL SENSE. IT IS TIME ASYMMETRY EMBEDDED IN NEUTRAL DESIGN.

By the late 2010s, this asymmetry began to surface indirectly in fertility clinic demographics. Clinics reported increasing numbers of patients who had spent extended periods in serial dating before attempting conception, often with little prior engagement with fertility literacy. This is not because information was unavailable, but because the systems governing early adult life did not prioritise it.

Digital dating platforms thus function as a temporal buffer between social pairing and biological consequence.

When that buffer collapses, users do not typically abandon optimisation logic. They transfer it. The same mindset that treated partner selection as a solvable algorithmic problem becomes receptive to medical and technological interventions promising to restore optionality.

This is the handoff point between platforms and clinics.

The monetisation logic of internet dating does not cause fertility decline. It conditions users to believe that time can always be managed later. When later arrives with less flexibility than expected, the search for technological compensation begins.

That search is not irrational. It is consistent.

Understanding this dynamic is essential before moving closer to the body. Fertility clinics inherit not only biological cases, but behavioural histories shaped by years of engagement with systems that optimised choice while obscuring cost.

The next layer of fertility technology does not replace dating platforms. It completes the logic they introduced.

References

Match Group Inc. Annual Reports and Form 10-K filings, 2019–2023.

Office for National Statistics (UK). Births in England and Wales statistical series.

Journal of Social and Personal Relationships. Studies on online dating duration and commitment timing.

Sociological Science. Research on partner search duration in digital environments.

United Nations Department of Economic and Social Affairs. World Fertility Data.

The Decline in Female Fertility

The natural decline in female fertility with age.

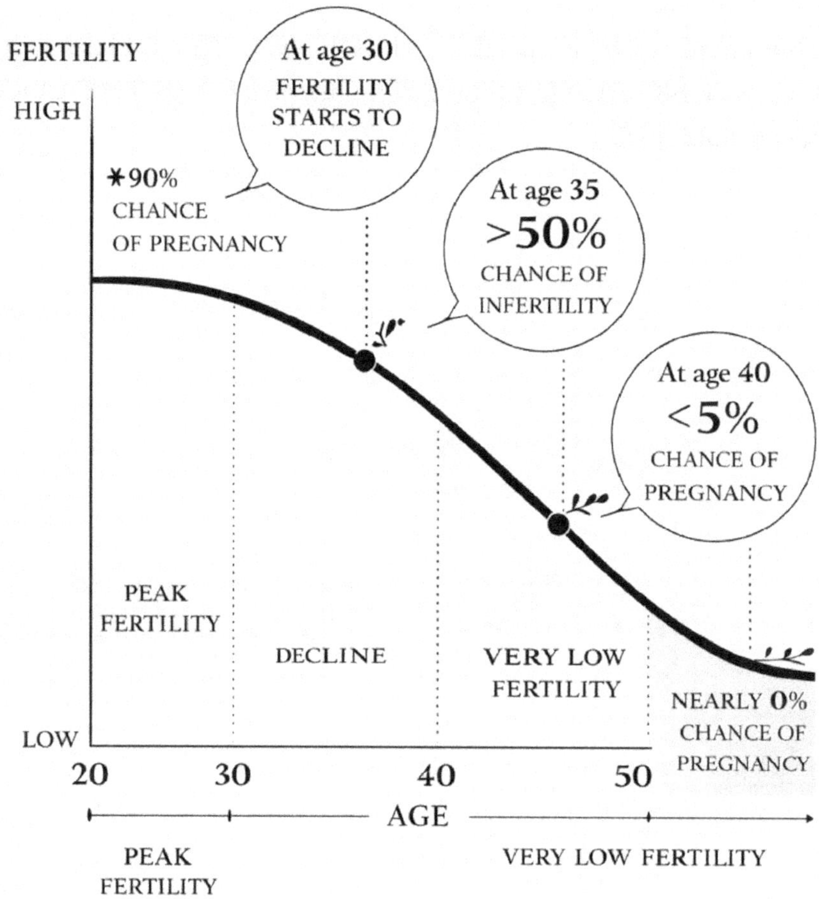

* APPROXIMATE AVERAGES

Chapter Three

Measuring the Body

Fertility Tracking, Diagnostics, and Biological Surveillance

3.1 Tracking, Self-Surveillance, and the Quantified Fertile Body

The transition from internet-mediated pairing to bodily measurement is not abrupt. It is a continuation of the same logic applied closer to the skin. Where dating platforms externalise selection and timing, fertility tracking internalises them. The body becomes observable, recordable, and comparable over time. Fertility moves from assumption to measurement.

This shift was not driven initially by medicine, but by consumer technology.

The modern fertility tracking ecosystem emerged from a convergence of smartphones, wearable sensors, and behavioural analytics in the late 2000s and early 2010s. Menstrual tracking apps were among the earliest health-focused applications to achieve mass adoption. Their appeal was simple: they promised foresight. They framed the reproductive cycle as something that could be anticipated rather than endured.

One of the most prominent examples is Flo Health, which reports over 230 million lifetime downloads globally by the early 2020s, with tens of millions of monthly active users. Flo is not unique.

Competing platforms such as Clue, Natural Cycles, and Glow collectively account for a substantial proportion of reproductive-age women in high-income countries engaging in some form of cycle tracking.

The scale of adoption matters because it signals a behavioural shift. Menstrual cycles, once tracked informally or not at all, are now routinely logged, analysed, and predicted. This practice introduces a new relationship between individuals and their reproductive biology. The cycle becomes a data series rather than a background rhythm.

From a fertility perspective, this is a critical transition.

Tracking systems typically begin with benign goals: understanding cycle length, predicting menstruation, or avoiding pregnancy. However, the same datasets underpin fertility prediction. Ovulation windows, luteal phase length, basal body temperature trends, and symptom correlations are extracted from repeated user inputs. Over time, the platform learns the user. The user, in turn, learns to interpret their body through the platform.

This feedback loop alters perception.

Where fertility was once something assumed to be present until proven otherwise, tracking reframes it as something that must be continually confirmed. Regularity becomes reassurance. Irregularity becomes a source of concern, even when

CLINICALLY INSIGNIFICANT. VARIABILITY, WHICH IS BIOLOGICALLY NORMAL, IS INCREASINGLY INTERPRETED THROUGH PROBABILISTIC LANGUAGE.

THE INTRODUCTION OF WEARABLE DEVICES INTENSIFIED THIS EFFECT.

DEVICES SUCH AS OURA AND LATER GENERATIONS OF SMARTWATCHES ADDED CONTINUOUS PHYSIOLOGICAL MONITORING TO THE REPRODUCTIVE PICTURE. BASAL BODY TEMPERATURE, HEART RATE VARIABILITY, SLEEP QUALITY, AND ACTIVITY LEVELS COULD NOW BE CORRELATED WITH CYCLE PHASES. THE BODY WAS NO LONGER SAMPLED PERIODICALLY. IT WAS OBSERVED CONTINUOUSLY.

FROM A SYSTEMS PERSPECTIVE, THIS MARKS THE POINT AT WHICH FERTILITY BECOMES PART OF THE BROADER QUANTIFIED-SELF MOVEMENT.

THE QUANTIFIED-SELF PARADIGM RESTS ON THE ASSUMPTION THAT INCREASED MEASUREMENT LEADS TO IMPROVED OUTCOMES. THIS ASSUMPTION HOLDS IN SOME DOMAINS. IN OTHERS, IT INTRODUCES NOISE, ANXIETY, AND FALSE PRECISION. FERTILITY TRACKING SITS UNEASILY BETWEEN THESE EXTREMES.

CLINICAL EVIDENCE SUPPORTS LIMITED ASPECTS OF TRACKING. BASAL BODY TEMPERATURE CHARTING AND LUTEINISING HORMONE DETECTION CAN HELP IDENTIFY OVULATION RETROSPECTIVELY OR IN NEAR REAL TIME. HOWEVER, POPULATION-LEVEL STUDIES PUBLISHED IN REPRODUCTIVE MEDICINE JOURNALS CONSISTENTLY SHOW WIDE INTER-INDIVIDUAL VARIABILITY. OVULATION DOES NOT ALWAYS OCCUR ON SCHEDULE. CYCLES SHIFT

Despite this, consumer platforms often present predictions with high confidence. The language of probability is softened into the language of certainty. "Fertile window" becomes a discrete interval rather than a distribution. This is not deception; it is interface simplification.

The effect on user behaviour is measurable.

Surveys conducted in the UK, EU, and United States during the late 2010s indicate that a significant proportion of users modify sexual behaviour based on app predictions. In contexts of conception avoidance, this can be effective. In contexts of conception seeking, it can introduce frustration when outcomes do not align with predictions. with stress, illness, travel, and age. Algorithms smooth these variations, but they cannot eliminate them.

From a fertility systems perspective, tracking technologies perform two functions simultaneously.

First, they increase reproductive literacy. Users gain vocabulary and awareness. Terms such as ovulation, luteal phase, and cycle length enter everyday conversation. This literacy is a precondition for later engagement with clinical fertility services.

Second, they shift responsibility inward. Fertility becomes something the individual monitors, manages, and optimises. Variations are experienced as personal signals rather than population-level statistics. When conception does not occur, the question "why?" arises earlier.

This early questioning is significant.

Historically, fertility concerns often emerged only after prolonged unsuccessful attempts at conception. With tracking, concern can arise before attempts even begin. Deviations from algorithmic expectations prompt self-investigation, online research, and eventually medical consultation.

In this way, tracking platforms act as early detection systems, not necessarily for pathology, but for perceived deviation from norm.

The demographic implications are subtle but important. As average age at first attempt increases, tolerance for uncertainty decreases. Tracking tools fill this gap by offering apparent control. They do not extend fertility, but they make its limits more visible.

This visibility interacts directly with time asymmetry.

For younger users, tracking often reinforces reassurance. For older users, it can accelerate anxiety. The same data, interpreted through different temporal lenses, produces different psychological effects. The technology does not adjust for this context. It delivers predictions uniformly.

By the early 2020s, regulatory attention began to follow adoption. Concerns around data privacy, medical claims, and predictive accuracy prompted scrutiny in multiple jurisdictions. Some platforms adjusted language, added disclaimers, or sought formal certification as medical devices. Others remained lifestyle products.

Regardless of regulatory classification, the functional role of tracking technologies remains consistent. They are the gateway layer between abstract fertility awareness and clinical intervention.

Once fertility is measured daily, it becomes difficult to return to assumption.

The quantified fertile body does not yet involve clinics, needles, or laboratories. It involves charts, notifications, and trend lines. But the logic is already medicalised. Variability is suspect. Prediction is expected. Outcomes are monitored.

This is the point at which fertility ceases to be a background condition and becomes an object of management.

The next layer extends this logic further, away from self-reporting and toward diagnostics that assign meaning to the data. Where tracking asks "when," diagnostics ask "why."

That transition marks the entry into formal reproductive medicine.

References

Flo Health. User statistics and platform disclosures, 2019–2023.

Office for National Statistics (UK). Conceptions and fertility indicators datasets.

Reproductive medicine literature on basal body temperature and ovulation prediction, including reviews in Human Reproduction and Fertility and Sterility.

Quantified-self and digital health studies published in Journal of Medical Internet Research.

3.2 Diagnostics, Male Visibility, and the Redistribution of Fertility Responsibility

The move from self-tracking to formal diagnostics marks a qualitative shift in how fertility is understood. Tracking produces patterns. Diagnostics assign meaning. Where apps ask when, diagnostics ask why. This transition pulls fertility out of the personal management layer and into institutional medicine.

Historically, this transition was asymmetrical.

For most of the twentieth century, fertility investigation focused overwhelmingly on women. This was not because female biology was uniquely complex, but because pregnancy occurs in the female body and because diagnostic access followed visibility. Male fertility, by contrast, was often treated as binary: either adequate or irrelevant. When conception failed, investigation defaulted toward the person whose body would carry the pregnancy.

That imbalance has begun to correct, not through ideology, but through instrumentation.

Modern fertility diagnostics now identify male factors in a substantial proportion of infertility cases. Large epidemiological reviews published in Human Reproduction and Fertility and Sterility consistently report that male factors contribute to approximately 40–50 percent of infertility cases, either as a primary or contributing cause. This statistic alone destabilises older assumptions.

The diagnostic shift is enabled by three developments: standardisation, automation, and consumer accessibility.

The foundational male fertility diagnostic remains semen analysis. Traditional semen analysis evaluates sperm concentration, motility, morphology, and volume against reference ranges defined by the World Health Organization. For decades, this test was underutilised, often delayed until after invasive female investigations had already occurred.

In the 2010s, this pattern began to change.

Advances in laboratory automation and imaging reduced variability in semen analysis and lowered costs. At the same time, consumer-facing diagnostic companies emerged, reframing male fertility testing as routine rather than exceptional. Companies such as Legacy and ExSeed Health introduced mail-in semen testing kits, allowing samples to be analysed outside traditional clinic settings.

The impact of this shift is not merely logistical. It is conceptual.

When male fertility becomes measurable, it becomes discussable. When it becomes discussable, responsibility redistributes. Fertility ceases to be a problem that happens to women and becomes a shared biological constraint.

This redistribution has several downstream effects.

First, it alters the sequencing of investigation. Clinical guidelines in many countries now recommend early or simultaneous assessment of both partners. This reduces unnecessary invasive procedures for women and shortens time to meaningful intervention. It also reframes infertility as a couple-level phenomenon rather than an individual pathology.

Second, it changes how failure is interpreted. When male parameters fall outside reference ranges, infertility is no longer explained away as bad timing or female irregularity. It becomes a diagnosable condition with identifiable variables. This does not guarantee resolution, but it removes ambiguity.

Third, it introduces male ageing into fertility discourse.

While male fertility declines more gradually than female fertility, it does decline. Sperm quality, DNA fragmentation rates, and epigenetic integrity change with age. Large cohort studies published in the British Medical Journal and Nature Reviews Urology have linked advanced paternal age to increased risks of miscarriage, autism spectrum disorders, and certain congenital anomalies. These risks remain low in absolute terms, but they are not zero.

Diagnostics make these gradients visible.

At the same time, female fertility diagnostics expanded in scope and precision. Hormonal panels measuring follicle-stimulating hormone, luteinising hormone, estradiol, and anti-Müllerian hormone became routine. Ultrasound-based antral follicle counts provided structural correlates of ovarian reserve. These metrics do not predict conception directly, but they constrain expectations.

The introduction of anti-Müllerian hormone testing in particular reshaped fertility conversations. AMH offered a quantifiable proxy for ovarian reserve, decoupled from cycle timing. It allowed clinicians to estimate remaining egg quantity, though not egg quality. This distinction is frequently misunderstood.

From a systems perspective, AMH functions as a time-compression signal. A low result does not preclude pregnancy, but it narrows the perceived window. It accelerates decision-making. It moves fertility from abstract possibility to bounded resource.

The combination of male diagnostics and expanded female testing produces a new equilibrium. Fertility becomes a joint, data-driven problem. However, this equilibrium is unevenly experienced.

Female diagnostics often precede invasive intervention. Blood draws, ultrasounds, and eventually procedures fall disproportionately on women, even when male factors are identified. Redistribution of responsibility does not equate to redistribution of burden.

This asymmetry is not primarily ethical. It is anatomical.

What diagnostics do accomplish is the removal of ambiguity. The era in which infertility could be indefinitely attributed to stress, lifestyle, or bad luck is closing. Diagnostic categories, while imperfect, create a sense of finitude. Something is measured. Something is named.

This naming has psychological consequences.

Patients encountering diagnostic thresholds often experience a shift from hope to strategy. Fertility becomes something to manage rather than assume. This mindset aligns naturally with technological intervention. Once a problem is quantified, optimisation becomes plausible.

From a demographic perspective, diagnostics also reveal population-level trends. Large-scale analyses of semen quality over multiple decades

SUGGEST DECLINING AVERAGE SPERM COUNTS IN WESTERN COUNTRIES, THOUGH THE MAGNITUDE AND INTERPRETATION OF THIS DECLINE REMAIN CONTESTED. REGARDLESS OF CAUSALITY, THE PERCEPTION OF DECLINE CONTRIBUTES TO DEMAND FOR TESTING AND INTERVENTION.

THE DIAGNOSTIC LAYER THUS SERVES AS A BRIDGE.

UPSTREAM, SELF-TRACKING PRODUCES AWARENESS WITHOUT AUTHORITY. DOWNSTREAM, CLINICAL INTERVENTION REQUIRES JUSTIFICATION. DIAGNOSTICS SUPPLY THAT JUSTIFICATION. THEY TRANSLATE SUBJECTIVE CONCERN INTO INSTITUTIONAL ACTION.

THEY ALSO PREPARE PATIENTS FOR ESCALATION.

ONCE FERTILITY IS FRAMED IN TERMS OF PARAMETERS, THRESHOLDS, AND PROBABILITIES, THE LEAP TO ASSISTED REPRODUCTION BECOMES CONCEPTUALLY SMALLER. THE LANGUAGE IS ALREADY MEDICAL. THE BODY IS ALREADY DATAFIED. INTERVENTION FEELS LIKE A CONTINUATION RATHER THAN A RUPTURE.

THIS IS WHY DIAGNOSTICS ARE A PIVOTAL LAYER IN THE FERTILITY TECHNOLOGY STACK. THEY REDISTRIBUTE RESPONSIBILITY, EXPOSE ASYMMETRIES, AND COMPRESS TIMELINES. THEY ALSO NORMALISE THE IDEA THAT REPRODUCTION CAN AND SHOULD BE MANAGED WITH TOOLS.

THE NEXT STEP MOVES BEYOND EXPLANATION INTO ACTION. WHERE DIAGNOSTICS DESCRIBE CONSTRAINTS, INTERVENTION ATTEMPTS TO OVERCOME THEM.

That is where laboratories, protocols, and industrial processes enter the story.

References

World Health Organization. WHO Laboratory Manual for the Examination and Processing of Human Semen, 5th and 6th editions.

Human Reproduction and Fertility and Sterility. Reviews on male-factor infertility and diagnostic contribution rates.

British Medical Journal. Studies on paternal age and reproductive outcomes.

Nature Reviews Urology. Reviews on sperm quality, DNA fragmentation, and ageing.

Clinical endocrinology literature on anti-Müllerian hormone and ovarian reserve assessment.

Chapter Four

When Probability Fails

Clinical Fertility Testing and the First Interventions

When the Numbers Arrived

The numbers didn't arrive dramatically.

They came in emails, in portals with sensible fonts, in conversations that were careful not to sound alarming. Nothing was wrong, exactly. Nothing was reassuring either. Everything sat just slightly off centre, like a picture that hasn't been hung properly.

Up until then, fertility had been something I thought about in terms of timing. Not dates, but seasons. Later. Soon. Eventually. Diagnostics don't work in that register. They don't care about intention. They care about ranges.

I remember noticing how quickly the language changed. Not in tone, but in grammar. Less future tense. More conditionals. Words like "within" and "relative" started appearing. Nobody said impossible. Nobody said urgent. Nobody needed to.

I became aware of my body in new, unglamorous ways. Not in pain, not in crisis. Just different. Blood tests meant bruises in places I wasn't used to looking at closely. Appointments rearranged

MY DAYS. I STARTED TIMING MEALS AROUND NOTHING IN PARTICULAR. MY HAIR FELT GREASIER THAN USUAL, WHICH I BLAMED ON HORMONES, THEN ON STRESS, THEN GAVE UP BLAMING AT ALL.

THE STRANGEST PART WAS HOW NORMAL EVERYTHING STILL LOOKED FROM THE OUTSIDE. I WAS HEALTHY. I SLEPT WELL. I WORKED. I LAUGHED AT THE RIGHT MOMENTS. INSIDE, THOUGH, SOMETHING HAD SHIFTED. FERTILITY HAD STOPPED BEING A BACKGROUND ASSUMPTION AND BECOME A SET OF VARIABLES I WAS NOW EXPECTED TO UNDERSTAND.

MALE FERTILITY ENTERING THE CONVERSATION CHANGED THINGS MORE THAN I EXPECTED. NOT EMOTIONALLY, EXACTLY. STRUCTURALLY. THE PROBLEM NO LONGER HAD A DEFAULT OWNER. RESPONSIBILITY REDISTRIBUTED ITSELF WITHOUT CEREMONY. THAT WAS A RELIEF, THOUGH NOT A SIMPLE ONE. SHARED RESPONSIBILITY DOESN'T MAKE DECISIONS EASIER. IT JUST MAKES THEM SHARED.

WHAT STAYED WITH ME WAS HOW QUIETLY THE DOOR CLOSED ON CERTAIN KINDS OF IGNORANCE. I COULD NO LONGER PRETEND NOT TO KNOW WHAT OVARIAN RESERVE MEANT, EVEN IF I DIDN'T WANT TO THINK ABOUT IT EVERY DAY. I COULDN'T UNKNOW THE EXISTENCE OF REFERENCE RANGES, EVEN IF THEY WEREN'T DESTINY. KNOWLEDGE DIDN'T ARRIVE WITH INSTRUCTIONS FOR HOW TO CARRY IT.

I STARTED READING MORE, AT FIRST FOR REASSURANCE, THEN FOR CONTEXT. THE EARLY ARTICLES WERE PRACTICAL. WHAT TESTS MEANT. WHAT THEY DIDN'T. WHERE VARIATION WAS NORMAL. LATER, THE READING DRIFTED. I NOTICED PATTERNS IN HOW FERTILITY WAS DISCUSSED, HOW RESPONSIBILITY WAS FRAMED, HOW QUICKLY OPTIMISATION LANGUAGE CREPT IN ONCE NUMBERS WERE INVOLVED.

THIS WASN'T PANIC. IT WAS ORIENTATION.

DIAGNOSTICS DIDN'T MAKE ANYTHING WORSE. THEY JUST MADE IT HARDER TO FLOAT. ONCE THE NUMBERS ARE ON THE TABLE, THE ROOM CHANGES. YOU SIT DIFFERENTLY. YOU LISTEN MORE CAREFULLY. YOU STOP ASSUMING TIME WILL SMOOTH THINGS OUT ON ITS OWN.

NOTHING HAD GONE WRONG YET. THAT WAS THE UNSETTLING PART. EVERYTHING WAS STILL POSSIBLE. IT WAS JUST NO LONGER ABSTRACT.

THAT'S WHEN I REALISED THAT UNDERSTANDING FERTILITY WASN'T GOING TO STOP AT UNDERSTANDING MY OWN.

CHAPTER FIVE

MANUFACTURED CONCEPTION

IVF, ICSI, AND THE INDUSTRIALISATION OF EARLY LIFE

5.1 IVF as Process, Not Miracle

In vitro fertilisation is often framed culturally as a last resort, a medical miracle, or a deeply personal intervention. Structurally, however, IVF is none of these things. It is an industrialised biological process designed to compensate for declining reproductive efficiency under conditions of time compression, diagnostic constraint, and probabilistic failure.

Understanding IVF properly requires stripping it of narrative exceptionalism and examining it as a production system.

Globally, IVF has moved from experimental procedure to routine clinical service in under half a century. Since the birth of the first IVF child in 1978, the technology has expanded rapidly. Estimates published in Human Reproduction and by the International Committee for Monitoring Assisted Reproductive Technologies suggest that over 8 million children worldwide have now been born via assisted reproductive technologies, with IVF accounting for the majority. In countries such as Denmark, Spain, and parts of Australia, between 6 and 10 percent of all births now involve assisted reproduction.

This scale alone necessitates standardisation.

Modern IVF clinics operate less like bespoke medical practices and more like tightly controlled production environments. While patient experience is framed around intimacy and care, the underlying structure is governed by protocols, throughput, failure rates, and optimisation loops. The biological variability of human reproduction is managed through redundancy rather than precision.

The core IVF workflow is broadly consistent across jurisdictions.

First, ovarian stimulation is induced pharmacologically. Gonadotropins are administered to override the body's natural cycle and produce multiple follicles simultaneously. This is not an enhancement of natural fertility, but a controlled disruption of it. The objective is not ovulation, but yield.

Second, oocyte retrieval is performed via transvaginal aspiration under ultrasound guidance. The procedure is standardised, time-bound, and repeatable. From an industrial perspective, this is the input stage. Eggs are harvested in batches, not individually.

Third, fertilisation occurs in the laboratory. This step may involve conventional IVF, where sperm and oocytes are co-incubated, or intracytoplasmic sperm injection, where a single sperm is mechanically injected into the oocyte. The choice reflects diagnostic inputs, not romance.

Fourth, embryos are cultured in controlled incubators for several days. Environmental parameters including temperature, pH, oxygen concentration, and nutrient media are tightly regulated. Development is monitored according to morphological criteria and, increasingly, time-lapse imaging.

FINALLY, ONE OR MORE EMBRYOS ARE TRANSFERRED, CRYOPRESERVED, OR DISCARDED ACCORDING TO CLINICAL STRATEGY, REGULATORY CONSTRAINTS, AND PATIENT CONSENT.

AT EVERY STAGE, LOSS IS EXPECTED.

THIS EXPECTATION DISTINGUISHES IVF FROM NATURAL CONCEPTION. NATURAL REPRODUCTION IS OPTIMISED FOR EFFICIENCY AT LOW YIELD. IVF IS OPTIMISED FOR YIELD UNDER INEFFICIENCY. THE PROCESS ASSUMES THAT MOST EGGS WILL NOT FERTILISE, MOST EMBRYOS WILL NOT DEVELOP NORMALLY, AND MOST TRANSFERS WILL NOT RESULT IN LIVE BIRTH. SUCCESS EMERGES STATISTICALLY, NOT DETERMINISTICALLY.

CLINIC PERFORMANCE METRICS REFLECT THIS REALITY.

ACCORDING TO DATA PUBLISHED BY THE UK HUMAN FERTILISATION AND EMBRYOLOGY AUTHORITY, LIVE BIRTH RATES PER EMBRYO TRANSFER VARY SHARPLY BY MATERNAL AGE. FOR WOMEN UNDER 35, RATES MAY EXCEED 30 PERCENT PER TRANSFER. BY AGE 40, THIS FALLS BELOW 15 PERCENT, AND BY 43, INTO SINGLE DIGITS. THESE FIGURES ARE NOT FAILURES OF TECHNIQUE. THEY ARE EXPRESSIONS OF BIOLOGY UNDER CONSTRAINT.

The industrial response to declining efficiency is not to slow down, but to scale.

This scaling is supported by a specialised supply chain. Companies such as Vitrolife and CooperSurgical provide culture media, incubators, micromanipulation tools, and consumables designed to reduce variance and increase reproducibility. These companies do not sell miracles. They sell reliability.

Pharmaceutical suppliers, including Merck KGaA, provide stimulation drugs calibrated for predictable ovarian response. Protocols are refined continuously based on population data rather than individual optimisation. Outliers are expected. The system absorbs them.

Automation has become increasingly central.

Time-lapse incubators allow continuous embryo monitoring without environmental disruption. Digital record systems track outcomes across thousands of cycles. Protocol adjustments are informed by aggregated results rather than anecdote. IVF clinics now operate with feedback loops characteristic of manufacturing environments.

This industrialisation has ethical and psychological consequences, but its origin is pragmatic.

IVF exists because spontaneous conception declines faster than desire for children. It compensates for delayed pairing, age-related fertility loss, and male-factor decline by concentrating reproductive effort into controlled windows. It is a technology of compression.

Critically, IVF does not replace natural fertility. It front-loads it. Eggs that might have been ovulated one per month are harvested in clusters. Fertilisation that might have occurred over years is attempted within days. Embryonic development that would have failed invisibly in utero now fails visibly in vitro.

Visibility changes perception.

For patients, IVF often feels emotionally intense because loss is observed rather than abstract. For clinics, loss is expected, logged, and factored into performance metrics. This asymmetry between lived experience and system design is a defining feature of industrial reproduction.

It is also why IVF feels simultaneously intimate and impersonal.

From a systems perspective, IVF is not controversial because it is unnatural. It is controversial because it exposes how inefficient natural reproduction already is. By making failure observable, it removes the illusion that fertility was ever guaranteed.

This exposure sets the stage for further optimisation.

Once conception is externalised, questions follow naturally. Why not intervene earlier? Why not select better embryos? Why not automate judgement? Why not reduce randomness further?

Those questions do not arise from ideology. They arise from process visibility.

IVF is the hinge point where reproduction stops being private and becomes procedural. Everything that follows builds on this shift.

References

Human Fertilisation and Embryology Authority (UK). Fertility Treatment Statistics, 2010–2023.

International Committee for Monitoring Assisted Reproductive Technologies (ICMART). World ART reports.

Human Reproduction. Global reviews on IVF outcomes and utilisation rates.

Fertility and Sterility. Clinical studies on age-stratified IVF success rates.

Vitrolife AB. Product and clinical background documentation.

CooperSurgical Inc. Reproductive medicine product literature.

Merck KGaA. Fertility treatment pharmacology disclosures.

5.2 ICSI, Male-Factor Infertility, and the Mechanisation of Fertilisation

Intracytoplasmic sperm injection, commonly abbreviated as ICSI, represents one of the most consequential shifts in modern reproductive medicine. Where conventional IVF still relies on sperm competition and autonomous fertilisation within a laboratory environment, ICSI removes competition entirely. Fertilisation becomes a mechanical act performed by a trained embryologist using micromanipulation tools under a microscope.

This shift is not philosophical. It is technical, and it was driven by necessity.

ICSI was first successfully performed in the early 1990s by Gianpiero Palermo and colleagues, in response to severe male-factor infertility cases where sperm concentration, motility, or morphology made conventional fertilisation unreliable or impossible. The procedure involved selecting a single viable sperm and injecting it directly into the cytoplasm of the oocyte using a glass micropipette.

What began as a rescue technique rapidly became routine.

By the late 2010s, ICSI accounted for more than 60 percent of IVF cycles worldwide, according to data compiled by the International Committee for Monitoring Assisted Reproductive Technologies

AND REPORTED IN HUMAN REPRODUCTION. IN SOME REGIONS, INCLUDING PARTS OF SOUTHERN EUROPE, ASIA, AND THE MIDDLE EAST, USAGE RATES EXCEED 80 PERCENT. THIS EXPANSION OCCURRED DESPITE ONGOING DEBATE ABOUT WHETHER ICSI IMPROVES OUTCOMES IN NON-MALE-FACTOR CASES.

THE REASONS FOR THIS EXPANSION ARE STRUCTURAL.

MALE-FACTOR INFERTILITY IS COMMON, HETEROGENEOUS, AND OFTEN POORLY REVERSIBLE. SEMEN QUALITY VARIES WIDELY ACROSS INDIVIDUALS AND OVER TIME. EVEN WITHIN A SINGLE EJACULATE, ONLY A FRACTION OF SPERM EXHIBIT PROGRESSIVE MOTILITY, NORMAL MORPHOLOGY, AND INTACT DNA. CONVENTIONAL IVF DEPENDS ON THIS MINORITY PERFORMING SUCCESSFULLY UNDER ARTIFICIAL CONDITIONS.

ICSI BYPASSES THIS VARIABILITY.

FROM AN INDUSTRIAL PERSPECTIVE, ICSI INCREASES PREDICTABILITY. IT REDUCES FERTILISATION FAILURE RATES, PARTICULARLY IN CASES OF LOW SPERM COUNT, POOR MOTILITY, OBSTRUCTIVE AZOOSPERMIA, OR PRIOR FAILED FERTILISATION. CLINICS FAVOUR PREDICTABILITY BECAUSE FAILED FERTILISATION REPRESENTS A COMPLETE CYCLE LOSS, WHEREAS LATER-STAGE FAILURE CAN STILL YIELD DATA AND INFORM ADJUSTMENT.

This preference has consequences.

ICSI shifts fertilisation from a probabilistic biological event to a manual selection process. An embryologist evaluates sperm visually and selects one based on motility and morphology. This selection is crude compared to natural processes, which involve biochemical signalling, competition, and multi-stage filtering within the female reproductive tract.

However, crude does not mean ineffective.

Large meta-analyses published in Fertility and Sterility indicate that ICSI significantly reduces the risk of total fertilisation failure compared to conventional IVF, particularly in male-factor cases. Live birth rates per cycle, however, are broadly similar when adjusted for indication and maternal age. This distinction is important. ICSI improves reliability, not ultimate success.

The widespread adoption of ICSI reflects clinic incentives rather than biological superiority.

From a systems perspective, ICSI reduces uncertainty at a critical bottleneck. Fertilisation is no longer left to chance. It is scheduled, logged, and reproducible. This aligns with industrial logic.

The mechanisation of fertilisation also reframes male infertility.

Historically, severe male infertility often led to donor sperm use or childlessness. ICSI enables biological parenthood even in cases where sperm are immotile, scarce, or surgically retrieved from testicular tissue. Techniques such as testicular sperm extraction and percutaneous epididymal sperm aspiration are now routinely paired with ICSI.

This capability expands the addressable population for IVF.

However, it also introduces new layers of risk and uncertainty.

By bypassing natural sperm selection mechanisms, ICSI may allow fertilisation by sperm that would not succeed in vivo. Concerns have been raised about potential increases in chromosomal abnormalities, imprinting disorders, and epigenetic effects. Large cohort studies have generally found that absolute risks remain low, but not zero.

The medical literature reflects this ambivalence.

Reviews in Human Reproduction Update and The Lancet note small but statistically significant increases in certain rare outcomes among ICSI-conceived children compared to spontaneous conception. These findings are difficult to interpret due to confounding by parental infertility factors. Nevertheless, they underscore that ICSI trades biological filtering for mechanical control.

Clinically, this trade-off is accepted.

From a fertility systems perspective, ICSI marks a decisive step toward engineering reproduction. The act of fertilisation is no longer emergent. It is executed.

This execution requires infrastructure.

ICSI depends on high-precision micromanipulation equipment, sterile consumables, and highly trained embryologists. Suppliers such as CooperSurgical and Vitrolife provide micromanipulators, injection pipettes, holding pipettes, and culture systems optimised for ICSI workflows.

Training is non-trivial. Embryologists must acquire fine motor skills comparable to microsurgeons. Variability between operators can affect outcomes. Clinics manage this through protocol standardisation, continuous training, and internal auditing.

ICSI thus accelerates the professionalisation of laboratory reproduction.

It also alters patient experience.

Where conventional IVF retains a conceptual link to natural fertilisation, ICSI makes the artificial nature of the process explicit. Fertilisation occurs not through interaction, but through intervention. For some patients, this clarity is reassuring. For others, it is unsettling.

From an ethical standpoint, ICSI is rarely controversial because it addresses a clear problem. Few argue that severe male-factor infertility should be left untreated on principle. The controversy arises instead from indication creep.

As ICSI becomes routine, its use expands beyond clear necessity. Clinics may prefer ICSI even in cases of unexplained infertility or advanced maternal age to reduce fertilisation failure risk. This normalises mechanical fertilisation as default rather than exception.

Once default shifts, expectations follow.

Patients begin to assume that fertilisation can be guaranteed if sufficient intervention is applied. This assumption is incorrect, but it is understandable within an industrial system that prioritises reliability.

ICSI also sets a precedent.

If fertilisation can be mechanised, why not selection? If selection can be guided, why not optimised? The logic of intervention does not stop at sperm injection. It propagates downstream.

In this sense, ICSI is not merely a technique. It is a philosophical inflection point embedded in practice. It reframes reproduction as a sequence of solvable sub-problems rather than a single uncertain event.

That reframing makes the next layer of fertility technology almost inevitable.

Once embryos exist outside the body, observable and enumerable, the question is no longer whether to intervene, but how far.

References (on page)

Palermo, G. et al. Early clinical reports on intracytoplasmic sperm injection, 1992–1994.

Human Reproduction. Global usage statistics and outcome analyses of ICSI.

Fertility and Sterility. Meta-analyses comparing IVF and ICSI fertilisation outcomes.

Human Reproduction Update. Reviews on safety, epigenetic considerations, and long-term outcomes of ICSI.

The Lancet. Population studies on assisted reproduction and child health outcomes.

International Committee for Monitoring Assisted Reproductive Technologies (ICMART). ART utilisation data.

CooperSurgical Inc. Micromanipulation and ICSI equipment documentation.

Vitrolife AB. Laboratory systems and embryo handling protocols.

5.3 Embryo Culture, Observation, and the Visibility of Failure

Once fertilisation has occurred, the centre of gravity in assisted reproduction shifts decisively from intervention to observation. Embryo culture is the phase in which IVF becomes most revealing, not because of what succeeds, but because of what fails. Processes that would normally unfold invisibly within the body are externalised, monitored, recorded, and judged in real time.

This visibility is not incidental. It is the defining feature of laboratory reproduction.

Following fertilisation, embryos are cultured in vitro for several days, typically to the cleavage stage on day two or three, or to the blastocyst stage on day five or six. During this period, embryos are maintained in incubators that tightly regulate temperature, gas composition, and humidity. Modern incubators maintain temperatures near 37°C, oxygen concentrations reduced to approximately 5 percent to mimic the uterine environment, and precisely buffered culture media designed to support early development.

The goal of culture is not growth alone. It is discrimination.

In natural conception, the majority of fertilised embryos fail to implant or arrest development without detection. Estimates published in Human

REPRODUCTION AND THE NEW ENGLAND JOURNAL OF MEDICINE SUGGEST THAT 30 TO 50 PERCENT OF FERTILISED HUMAN EMBRYOS FAIL PRIOR TO IMPLANTATION, OFTEN DUE TO CHROMOSOMAL ABNORMALITIES INCOMPATIBLE WITH DEVELOPMENT. THESE LOSSES ARE LARGELY INVISIBLE.

IN IVF, THEY ARE NOT.

EMBRYOS ARE GRADED ACCORDING TO MORPHOLOGY, CLEAVAGE TIMING, SYMMETRY, FRAGMENTATION, AND LATER, BLASTOCYST EXPANSION AND INNER CELL MASS QUALITY. THESE CRITERIA ARE CODIFIED IN SCORING SYSTEMS USED GLOBALLY, ALLOWING EMBRYOLOGISTS TO RANK EMBRYOS ACCORDING TO DEVELOPMENTAL POTENTIAL. THE SCORING SYSTEMS ARE PROBABILISTIC. A HIGH-GRADE EMBRYO MAY STILL FAIL. A LOW-GRADE EMBRYO MAY STILL RESULT IN LIVE BIRTH. HOWEVER, OUTCOMES CORRELATE STRONGLY ENOUGH WITH GRADING TO JUSTIFY ITS USE.

THE INDUSTRIAL LOGIC IS STRAIGHTFORWARD.

GIVEN FINITE TRANSFER OPPORTUNITIES AND DECLINING SUCCESS RATES WITH AGE, CLINICS AIM TO MAXIMISE THE LIKELIHOOD OF IMPLANTATION PER TRANSFER. THIS NECESSITATES RANKING. RANKING NECESSITATES OBSERVATION. OBSERVATION NECESSITATES ACCEPTING THAT SOME EMBRYOS WILL BE DEEMED INFERIOR AND EXCLUDED FROM FURTHER USE.

The psychological impact of this process is profound for patients, but structurally it is routine.

The introduction of extended culture to the blastocyst stage intensified this dynamic. Early IVF protocols often transferred embryos at day two or three, before significant differentiation had occurred. Advances in culture media and incubator stability enabled routine culture to day five or six, allowing embryos with limited developmental competence to arrest before transfer.

From a clinical perspective, this improves efficiency. Blastocyst transfer is associated with higher implantation rates per transfer and lower rates of multiple pregnancy when single embryos are transferred. From a systems perspective, it concentrates success into fewer attempts.

However, extended culture also increases attrition. Many embryos that appear viable at day three fail to reach blastocyst stage. This is not a failure of technique. It is a revelation of underlying developmental limitation.

Time-lapse imaging systems further increase visibility.

Technologies such as continuous embryo monitoring allow embryologists to observe developmental kinetics without removing embryos from incubators. Time-lapse systems generate detailed datasets on cleavage timing, cell division patterns, and morphological changes. These data are

INCREASINGLY INTEGRATED INTO DECISION-MAKING, SOMETIMES AUGMENTED BY ALGORITHMIC ANALYSIS.

SUPPLIERS SUCH AS VITROLIFE HAVE COMMERCIALISED INTEGRATED INCUBATOR AND IMAGING PLATFORMS THAT STANDARDISE OBSERVATION ACROSS CLINICS. THESE SYSTEMS REDUCE ENVIRONMENTAL DISTURBANCE AND INCREASE DATA DENSITY. THEY ALSO SHIFT EMBRYOLOGY FROM EPISODIC ASSESSMENT TO CONTINUOUS SURVEILLANCE.

THIS SURVEILLANCE CHANGES HOW FAILURE IS UNDERSTOOD.

IN NATURAL REPRODUCTION, FAILURE IS ABSTRACT. IN IVF, FAILURE IS VISUAL, TIMESTAMPED, AND ATTRIBUTABLE. AN EMBRYO ARRESTS AT THE EIGHT-CELL STAGE. A BLASTOCYST FAILS TO EXPAND. AN INNER CELL MASS APPEARS POOR. EACH EVENT IS RECORDED. EACH LOSS IS CONTEXTUALISED.

FROM A SYSTEMS PERSPECTIVE, THIS VISIBILITY ENABLES LEARNING. CLINICS AGGREGATE OUTCOMES ACROSS THOUSANDS OF CYCLES TO REFINE PROTOCOLS, ADJUST STIMULATION STRATEGIES, AND BENCHMARK PERFORMANCE. FROM AN INDIVIDUAL PERSPECTIVE, IT TRANSFORMS LOSS INTO A SERIES OF OBSERVABLE EVENTS RATHER THAN AN ABSENCE OF OUTCOME.

THIS TRANSFORMATION HAS ETHICAL IMPLICATIONS, BUT IT ORIGINATES IN PROCESS OPTIMISATION.

The logic of embryo culture is not to eliminate failure. It is to identify it earlier. By front-loading attrition into the laboratory phase, IVF reduces the likelihood of failed implantation or early miscarriage after transfer. This is often presented as compassionate. It is also efficient.

Selection pressure is thus applied before pregnancy begins.

The question of how far this selection should extend is unavoidable.

Once embryos are cultured, graded, and ranked, it becomes technically feasible to add additional layers of discrimination. Morphology can be supplemented with genetic screening. Developmental kinetics can be combined with chromosomal analysis. The embryo becomes a dataset rather than a singular possibility.

It is important to note that embryo grading and culture do not guarantee success. Live birth rates remain constrained by maternal age, uterine receptivity, and stochastic factors not captured by laboratory metrics. However, grading improves average outcomes enough to justify its universal adoption.

From an industrial standpoint, embryo culture is the quality control stage of reproduction.

It is here that IVF most clearly departs from the language of hope and enters the language of yield. Embryos are inputs, outputs, and intermediates. Losses are expected. Improvements are incremental.

This framing does not negate the emotional weight of the process, but it explains its structure.

By the time embryos reach the point of transfer or cryopreservation, they have already passed through multiple filters. Each filter narrows the field. Each narrowing is justified by probability.

This narrowing prepares the ground for the next step.

If embryos can be observed and ranked, they can also be analysed. If they can be analysed, they can be selected not just for viability, but for genetic composition. The move from observation to screening is therefore not a leap, but a continuation.

The laboratory has already made the invisible visible. Screening simply assigns names to what is already seen.

References (on page)

Human Reproduction. Reviews on embryo development, implantation failure, and IVF attrition rates.

The New England Journal of Medicine. Studies on early embryonic loss and chromosomal abnormality prevalence.

Fertility and Sterility. Research on blastocyst culture outcomes and single embryo transfer.

Vitrolife AB. Embryo culture systems, time-lapse imaging platforms, and laboratory protocols.

Human Reproduction Update. Meta-analyses on embryo grading systems and predictive value.

International Committee for Monitoring Assisted Reproductive Technologies (ICMART). ART outcome statistics.

5.4 Genetic Screening, Selection Logic, and the Soft Edge of Eugenics

Genetic screening of embryos represents the point at which assisted reproduction crosses from optimisation of process into optimisation of outcome. Up to this stage, IVF interventions aim to increase the probability that conception will occur and that development will proceed. With genetic screening, the intervention shifts toward deciding which potential lives are allowed to proceed at all.

This transition is neither abrupt nor explicitly ideological. It is procedural.

Preimplantation genetic testing, most commonly preimplantation genetic testing for aneuploidy (PGT-A), was developed to reduce implantation failure and miscarriage by identifying embryos with abnormal chromosome numbers. Aneuploidy increases sharply with maternal age and is a leading cause of early pregnancy loss. From a clinical perspective, screening offers a way to avoid transfers that are statistically unlikely to succeed.

The technical workflow is now well established.

After embryos reach the blastocyst stage, a small number of cells are biopsied from the trophectoderm, the outer cell layer that will later form the placenta. These cells are analysed using genomic technologies capable of detecting whole-chromosome gains and losses. The embryo itself is typically cryopreserved while results are processed.

This process relies on an industrial genomics supply chain.

Sequencing and array technologies provided by companies such as Illumina underpin much of modern preimplantation testing. Clinical testing services are offered by specialised laboratories, including Natera, which markets PGT-A and related reproductive genomic tests to fertility clinics worldwide. These companies do not operate clinics. They operate analytical infrastructure.

The results of PGT-A are reported categorically.

Embryos are typically classified as euploid (chromosomally normal), aneuploid (abnormal), or mosaic (containing both normal and abnormal cell lines). Euploid embryos are prioritised for transfer. Aneuploid embryos are usually excluded. Mosaic embryos occupy a contested middle ground.

From a systems perspective, this classification transforms reproduction into a triage process.

Selection logic is explicit. Embryos are not transferred in the order they are created, but in the order they score. Age, morphology, developmental timing, and now genetic composition are combined into a ranked list. The goal is not perfection. It is risk reduction.

Clinical data support this approach in narrow terms.

Meta-analyses published in Human Reproduction and Fertility and Sterility indicate that PGT-A can reduce miscarriage rates and improve implantation rates per transfer in certain populations, particularly women of advanced maternal age or those with recurrent pregnancy loss. However, evidence that PGT-A improves cumulative live birth rates per IVF cycle remains mixed. Screening does not create more viable embryos. It redistributes which embryos are attempted first.

This distinction is critical.

PGT-A does not increase fertility. It increases efficiency under constraint.

THE EXPANSION OF GENETIC SCREENING REFLECTS THIS EFFICIENCY LOGIC. AS IVF CYCLES BECOME MORE EXPENSIVE, EMOTIONALLY TAXING, AND TIME-LIMITED, REDUCING FAILED TRANSFERS BECOMES ECONOMICALLY AND PSYCHOLOGICALLY ATTRACTIVE. SCREENING PROMISES FEWER FALSE STARTS.

HOWEVER, SCREENING INTRODUCES NEW FORMS OF LOSS.

EMBRYOS CLASSIFIED AS ANEUPLOID ARE TYPICALLY DISCARDED OR INDEFINITELY STORED. SOME OF THESE EMBRYOS MAY HAVE DEVELOPED INTO HEALTHY PREGNANCIES IF TRANSFERRED, PARTICULARLY GIVEN THE LIMITATIONS OF BIOPSY SAMPLING AND THE PHENOMENON OF MOSAICISM. THE FALSE-NEGATIVE AND FALSE-POSITIVE RATES OF PGT-A REMAIN AREAS OF ACTIVE RESEARCH AND DEBATE.

THE SYSTEM ACCEPTS THIS UNCERTAINTY.

FROM AN INDUSTRIAL STANDPOINT, FALSE EXCLUSIONS ARE AN ACCEPTABLE TRADE-OFF FOR REDUCED DOWNSTREAM FAILURE. THIS IS A FAMILIAR CALCULUS IN QUALITY CONTROL SYSTEMS. IN MANUFACTURING, REJECTING BORDERLINE UNITS REDUCES CUSTOMER RETURNS EVEN IF SOME FUNCTIONAL PRODUCTS ARE DISCARDED.

REPRODUCTIVE SCREENING INHERITS THIS LOGIC.

THE ETHICAL WEIGHT OF THIS INHERITANCE IS SOFTENED BY LANGUAGE. SCREENING IS FRAMED AS INFORMATION. SELECTION IS FRAMED AS CHOICE. CLINICS EMPHASISE PATIENT AUTONOMY. THE SYSTEM RARELY USES THE WORD EXCLUSION, THOUGH THAT IS WHAT OCCURS.

Importantly, PGT-A is generally limited to chromosomal number rather than specific traits. This boundary is often cited as a safeguard against eugenics. However, the boundary is procedural, not conceptual.

Once embryos are screened for viability, the rationale for screening expands naturally. Testing for monogenic disorders (PGT-M) is already routine for families with known heritable diseases. Sex selection for medical reasons is permitted in many jurisdictions. Non-medical sex selection is restricted in some countries and permitted in others.

Each extension is justified individually.

What emerges is not a programmatic eugenics project, but a consumer-driven selection gradient. No authority mandates which embryos are preferred. Preferences accumulate through clinical practice, patient demand, and market availability.

This is what makes modern reproductive selection distinct from historical eugenics. There is no central ideology. There is no explicit social goal. There is only optimisation under scarcity.

Yet the effect is real.

Population-level modelling suggests that widespread embryo screening can subtly alter the distribution of genetic traits over time, even when limited to disease avoidance. The removal of embryos with certain chromosomal patterns reduces the frequency of those patterns in future populations. This is selection by omission rather than design.

The system does not ask whether this is desirable. It asks whether it is effective.

From a clinical perspective, genetic screening also reshapes patient expectations. Once embryos are classified, uncertainty feels less tolerable. Transferring a non-screened embryo can feel reckless rather than hopeful. The absence of information becomes a liability.

This psychological shift reinforces adoption.

Clinics that do not offer screening may appear outdated. Patients may feel they are forgoing an available advantage. Screening becomes default not because it is mandated, but because it aligns with a broader cultural intolerance for avoidable risk.

The boundary between medical necessity and elective optimisation thus erodes quietly.

It is important to emphasise that this erosion is incremental. There is no moment at which screening

BECOMES SINISTER. EACH STEP IS DEFENSIBLE IN ISOLATION. EACH IS GROUNDED IN REAL SUFFERING AVOIDED. MISCARRIAGES PREVENTED. FAILED CYCLES REDUCED.

THE CUMULATIVE EFFECT, HOWEVER, IS A REPRODUCTIVE SYSTEM INCREASINGLY ORIENTED TOWARD SELECTION BEFORE IMPLANTATION RATHER THAN ACCEPTANCE AFTER CONCEPTION.

THIS ORIENTATION HAS CONSEQUENCES FOR HOW REPRODUCTION IS UNDERSTOOD. LIFE BECOMES SOMETHING THAT MUST JUSTIFY ITS CONTINUATION BEFORE IT BEGINS. EMBRYOS ARE EVALUATED NOT ONLY FOR VIABILITY, BUT FOR DESIRABILITY UNDER DEFINED CRITERIA.

THE LANGUAGE OF CHOICE OBSCURES THE ASYMMETRY OF POWER.

EMBRYOS DO NOT CHOOSE. SYSTEMS CHOOSE ON BEHALF OF FUTURE OUTCOMES. PATIENTS PARTICIPATE, BUT WITHIN A MENU OF OPTIONS DEFINED BY TECHNOLOGY, REGULATION, AND MARKET FORCES.

AT THIS STAGE, THE LOGIC OF ASSISTED REPRODUCTION HAS MOVED DECISIVELY BEYOND ASSISTANCE. IT HAS ENTERED THE DOMAIN OF CURATION.

THE NEXT STEP FOLLOWS DIRECTLY FROM THIS LOGIC.

If embryos can be screened for chromosomal normality, they can be assessed for increasingly granular genetic information. If assessment becomes more granular, selection pressure increases. If selection pressure increases, reproduction becomes less about overcoming infertility and more about shaping outcomes.

The technologies that enable this shift are already emerging.

What remains is not a question of possibility, but of threshold.

References

Human Reproduction. Meta-analyses on PGT-A outcomes, miscarriage rates, and implantation efficiency.

Fertility and Sterility. Reviews on preimplantation genetic testing, mosaicism, and clinical utility.

The New England Journal of Medicine. Studies on chromosomal abnormalities and early pregnancy loss.

Illumina Inc. Genomic sequencing technologies used in reproductive medicine.

Natera Inc. Clinical documentation on PGT-A and reproductive genetic testing services.

Human Reproduction Update. Ethical and clinical discussions of embryo screening and selection.

International Committee for Monitoring Assisted Reproductive Technologies (ICMART). ART practice trends.

5.5 Mosaicism, Borderline Embryos, and Decision-Making Under Uncertainty

The introduction of genetic screening did not eliminate uncertainty from assisted reproduction. It reclassified it.

Where earlier IVF practice confronted uncertainty through repetition and probability, genetic screening confronts it through categorisation. Embryos are sorted into bins that imply action or exclusion. Mosaicism occupies the space where this logic fails to resolve cleanly. It is the category that exposes the limits of precision in reproductive genomics.

Mosaic embryos are those in which genetic testing identifies both euploid and aneuploid cell lines within the sampled tissue. In practical terms, this means that some cells appear chromosomally normal while others do not. The prevalence of mosaicism in blastocyst-stage embryos is not trivial. Large clinical datasets suggest that 10 to 30 percent of embryos tested via PGT-A may be classified as mosaic, depending on testing thresholds, laboratory protocols, and analytical platforms.

This classification emerged alongside improvements in genomic resolution.

Early preimplantation testing technologies lacked the sensitivity to detect mosaicism reliably. As next-generation sequencing and array-based methods became standard, previously invisible variation became visible. What was once labelled "normal" or "abnormal" acquired gradations.

Visibility created a problem.

Clinical outcomes associated with mosaic embryo transfer are heterogeneous. Published studies in Fertility and Sterility and Human Reproduction report that some mosaic embryos can implant and result in healthy live births, while others fail to implant or miscarry. Outcomes appear to correlate with the proportion of abnormal cells detected, the specific chromosomes involved, and maternal age, but predictive certainty remains limited.

From a systems perspective, mosaicism introduces decision-making under uncertainty into a process increasingly designed to minimise it.

Clinics respond by stratifying mosaic embryos further. Low-level mosaics may be considered for transfer under certain conditions. High-level mosaics are often deprioritised or excluded. Some clinics adopt conservative thresholds, others permissive ones. Regulatory guidance varies by country and evolves as evidence accumulates.

This variability matters.

Patients encountering mosaic results are presented not with clear recommendations, but with risk profiles. The language shifts again. Terms such as "potential," "chance," and "unknown" return to the conversation, but now embedded in genomic detail. Decision-making becomes more complex rather than simpler.

The burden of this complexity falls unevenly.

For patients with multiple euploid embryos, mosaic results are often academic. For patients with few embryos, particularly those of advanced maternal age or with diminished ovarian reserve, mosaic embryos may represent the only remaining possibility. The same classification carries radically different implications depending on context.

This contextual dependence exposes a structural tension.

Genetic screening was adopted to reduce failed transfers and miscarriages. Mosaicism reintroduces both risks in a conditional form. Transferring a mosaic embryo may result in failure, but not transferring it may eliminate the only chance of pregnancy. The system offers information without resolution.

From an industrial standpoint, mosaicism complicates optimisation. Quality control systems rely on clear accept/reject criteria. Borderline units undermine throughput efficiency. In manufacturing, such units are often discarded. In reproduction, discarding potential lives carries ethical weight that cannot be abstracted away.

Clinics manage this tension through counselling frameworks rather than algorithmic rules.

Genetic counsellors play a central role at this stage. They contextualise risk, explain limitations of testing, and help patients navigate trade-offs. The quality of counselling varies widely and is influenced by institutional culture, legal environment, and resource availability.

Importantly, counselling does not eliminate asymmetry.

Patients are asked to decide under conditions they cannot fully evaluate. Probabilities are presented, but lived consequences remain opaque. A ten percent risk of abnormal outcome is not experienced as ten percent. It is experienced as possibility.

This experiential gap is widened by the technical nature of mosaicism itself. Most patients lack intuitive frameworks for understanding chromosomal variation. Even clinicians acknowledge uncertainty. The literature evolves faster than consensus.

From a fertility systems perspective, mosaicism functions as a stress test.

It tests the assumption that more information always improves outcomes. It tests the belief that selection logic can be indefinitely refined. It tests the tolerance of patients and clinics for ambiguity in a system designed to minimise it.

The existence of mosaic embryos also reveals the limits of sampling.

PGT-A analyses a small number of cells from the trophectoderm. It does not sample the inner cell mass that will form the fetus. Mosaicism detected in the biopsy may not reflect the genetic status of the embryo as a whole. Conversely, undetected abnormalities may still exist.

This sampling limitation is well documented in the medical literature and acknowledged in professional guidelines. It is not a flaw of execution, but a constraint of method.

Despite this, classification persists because it offers relative ordering. Even imperfect information can be used to prioritise options under scarcity. The alternative is unranked choice, which feels riskier in a system that has already embraced optimisation.

Mosaicism therefore occupies an uneasy position.

It is neither a clear contraindication nor a clear endorsement. It resists clean integration into selection pipelines. It forces re-engagement with uncertainty at precisely the stage where patients expect clarity.

This re-engagement has downstream effects.

Some patients decline mosaic transfers categorically. Others accept them after euploid options are exhausted. Some clinics refuse to transfer certain mosaic types. Others specialise in them. The market fragments around risk tolerance.

This fragmentation reinforces inequality.

Patients with greater financial resources can pursue additional cycles to obtain euploid embryos, avoiding mosaic decisions entirely. Patients with fewer resources may face harder choices sooner. The technology does not create this inequality, but it makes it visible.

From an ethical standpoint, mosaicism challenges the implicit narrative of control that accompanies advanced reproductive technology. It reminds both clinicians and patients that biological systems remain noisy, probabilistic, and partially inscrutable.

Yet the system does not retreat.

Research continues to refine mosaic classification. Outcome registries grow. Algorithms are trained on larger datasets. The hope is that uncertainty can be reduced through scale.

Whether it can be eliminated remains doubtful.

Mosaicism thus marks the boundary where selection logic encounters irreducible complexity. It is not the end of screening, but it is the point at which screening reveals its own limits.

Beyond this boundary, the question shifts.

If uncertainty persists even with genomic insight, should selection become more granular, or should acceptance re-enter the system? The technologies that follow attempt the former. They seek to extract more signal from the embryo before implantation.

The next step does not abandon uncertainty. It attempts to outpace it.

References

Fertility and Sterility. Clinical outcome studies on mosaic embryo transfer.

Human Reproduction. Reviews on mosaicism prevalence, classification, and implantation outcomes.

Human Reproduction Update. Consensus statements and ethical analyses of mosaic embryo transfer.

Preimplantation genetic testing guidelines from professional reproductive medicine societies.

Clinical genomics literature on sampling limitations and interpretive uncertainty in PGT-A.

5.6 Polygenic Risk Scores, Trait Prediction, and the Expansion of Selection Criteria

If preimplantation genetic testing for aneuploidy introduced selection into assisted reproduction, polygenic risk scoring extends that selection from discrete abnormalities into gradients of probability. This extension does not represent a sudden ethical rupture. It represents an incremental broadening of the same logic already in place.

Polygenic risk scores, commonly abbreviated as PRS, estimate an individual's genetic predisposition to complex traits or diseases by aggregating the effects of many genetic variants across the genome. Unlike monogenic disorders, which are caused by mutations in a single gene, polygenic traits emerge from the interaction of hundreds or thousands of loci, each contributing a small effect.

The technology underlying PRS was not developed for reproduction. It emerged from population genomics and large-scale genome-wide association studies. As datasets expanded into the millions, correlations between genetic variants and traits such as height, body mass index, coronary artery disease, and type 2 diabetes became statistically robust enough to support individual-level risk estimation.

THE TRANSLATION OF THIS TECHNOLOGY INTO FERTILITY MEDICINE WAS A MATTER OF OPPORTUNITY RATHER THAN INTENT.

ONCE EMBRYOS ARE BIOPSIED, SEQUENCED, AND ANALYSED FOR CHROMOSOMAL STATUS, ADDITIONAL GENETIC INFORMATION IS ALREADY PRESENT. THE MARGINAL COST OF COMPUTING POLYGENIC SCORES FROM EXISTING GENOMIC DATA IS LOW. THE CONCEPTUAL BARRIER, RATHER THAN THE TECHNICAL ONE, HAS BEEN THE PRIMARY CONSTRAINT.

THAT BARRIER HAS BEGUN TO ERODE.

COMPANIES SUCH AS GENOMIC PREDICTION HAVE PIONEERED THE APPLICATION OF PRS TO EMBRYO SELECTION, MARKETING SERVICES THAT ASSESS EMBRYOS NOT ONLY FOR ANEUPLOIDY BUT FOR RELATIVE RISK OF COMMON DISEASES. THEIR APPROACH DOES NOT CLAIM CERTAINTY. IT OFFERS RANKING. EMBRYOS ARE SCORED RELATIVE TO ONE ANOTHER WITHIN A COHORT, ALLOWING PROSPECTIVE PARENTS TO SELECT EMBRYOS PREDICTED TO HAVE LOWER GENETIC RISK PROFILES.

THIS FRAMING IS CRITICAL.

POLYGENIC SCREENING DOES NOT PROMISE DISEASE-FREE OUTCOMES. IT PROMISES RISK REDUCTION ACROSS DISTRIBUTIONS. AN EMBRYO WITH A LOWER POLYGENIC RISK SCORE FOR CORONARY ARTERY DISEASE MAY STILL DEVELOP HEART DISEASE. AN EMBRYO WITH A HIGHER SCORE MAY NEVER DO SO. THE SELECTION OPERATES AT THE LEVEL OF PROBABILITY, NOT DESTINY.

From a systems perspective, this is consistent with existing IVF practice.

IVF already prioritises embryos based on probabilistic criteria. Morphology predicts implantation likelihood. Euploidy predicts reduced miscarriage risk. Polygenic scores extend this probabilistic triage into postnatal life.

The difference lies in temporal reach.

Where earlier selection criteria aim to improve the chance of pregnancy, polygenic selection aims to influence the future health profile of the resulting individual. The optimisation target shifts from reproductive success to life-course outcomes.

This shift has profound implications.

Polygenic scores are trait-specific and population-dependent. Their predictive accuracy varies widely depending on ancestry, dataset composition, and trait heritability. Scores derived from European-ancestry datasets perform poorly when applied to embryos with diverse genetic backgrounds. This limitation is well documented in genomics literature and acknowledged by practitioners.

Despite this, the appeal of PRS persists.

For families with a history of chronic disease, even modest risk reduction can feel meaningful. The logic mirrors earlier justifications for aneuploidy screening. Avoiding known suffering, even probabilistically, is framed as responsible.

Regulatory environments differ sharply.

In the United States, polygenic embryo screening is permitted under existing frameworks that emphasise patient autonomy and market choice. In many European countries, regulatory authorities restrict embryo selection to medical indications with clear clinical utility. Enforcement varies. Cross-border fertility tourism exploits these differences.

The absence of harmonised regulation accelerates diffusion.

Clinics operating in permissive jurisdictions gain competitive advantage by offering expanded screening menus. Patients dissatisfied with restrictions elsewhere travel. The technology spreads not because it is universally endorsed, but because it is available somewhere.

This diffusion reinforces a consumer-driven selection gradient.

Each incremental addition to screening criteria appears defensible in isolation. Avoiding severe disease. Reducing chronic illness risk. Improving long-term health. The cumulative effect, however, is a reproductive system increasingly oriented toward pre-emptive optimisation.

The term "soft eugenics" is sometimes applied to this dynamic, but it obscures more than it clarifies.

There is no coercion. No state mandate. No explicit hierarchy of human worth. What exists instead is a marketplace of probabilities in which selection emerges from aggregated individual choices rather than central planning.

This distinction matters legally and ethically, but it does not erase population-level effects.

Modelling studies in genetics suggest that even modest selection pressures applied consistently across generations can shift trait distributions. When selection is applied at the embryo stage, these effects operate more efficiently than postnatal interventions.

However, the magnitude of these shifts remains uncertain.

Polygenic traits are influenced by environment, behaviour, and stochastic factors that cannot be controlled genetically. The predictive power of PRS, while statistically significant, is far from absolute. Overconfidence in genetic determinism remains a risk.

Clinically, the introduction of PRS complicates counselling.

Patients are now asked to weigh abstract future risks against present-day embryos. The decision space expands dramatically. Instead of choosing between transfer or discard based on viability, patients choose between ranked futures based on statistical models.

This choice architecture favours those comfortable with probabilistic reasoning and long-term planning. Others experience decision paralysis. The system offers more information without necessarily improving clarity.

From an industrial perspective, PRS represent a scalable extension of existing genomics pipelines. The marginal cost is low. The data is already generated. The primary investment lies in analytics and interpretation.

This makes expansion likely.

As genomic databases grow and models improve, additional traits become scoreable. Cognitive performance, educational attainment, and behavioural traits are frequently cited in public discourse, though most clinics currently restrict screening to health-related endpoints. The boundary between health and performance, however, is not fixed.

Height, for example, is not a disease, yet short stature can have health and social correlates. Educational attainment is not pathology, yet it correlates with socioeconomic outcomes. Each potential trait sits on a spectrum between medical relevance and social desirability.

The system does not resolve this ambiguity. It manages it through policy, pricing, and consent.

Importantly, PRS do not replace earlier selection criteria. They layer on top of them. Aneuploidy screening remains foundational. Morphology still matters. Polygenic scores refine ranking rather than redefining viability.

This layering produces increasingly narrow funnels.

Embryos that pass chromosomal screening are ranked by morphology. Among those, polygenic scores may further differentiate. The number of embryos reaching transfer shrinks. Each step is justified by efficiency.

The effect is cumulative.

What began as a response to infertility evolves into a mechanism for anticipatory governance of life. Not through ideology, but through optimisation under constraint.

Whether this trajectory stabilises or accelerates depends less on ethical consensus than on technical capability and market demand. History suggests that once probabilistic optimisation is possible and affordable, it becomes difficult to abandon.

Polygenic embryo screening thus represents not the endpoint of reproductive selection, but a midpoint.

Beyond it lie technologies that do not merely rank embryos, but seek to generate them differently. If selection can only reduce risk within existing variation, modification promises to alter the variation itself.

THE NEXT STEP DOES NOT ASK WHICH EMBRYO IS BEST.

IT ASKS WHETHER EMBRYOS SHOULD BE MADE DIFFERENTLY IN THE FIRST PLACE.

REFERENCES

NATURE REVIEWS GENETICS. REVIEWS ON POLYGENIC RISK SCORES, PREDICTIVE POWER, AND LIMITATIONS.

HUMAN REPRODUCTION. DISCUSSIONS ON THE APPLICATION OF PRS IN ASSISTED REPRODUCTION.

FERTILITY AND STERILITY. ETHICAL AND CLINICAL ANALYSES OF EXPANDED EMBRYO SCREENING.

GENOMIC PREDICTION INC. PUBLIC DOCUMENTATION ON POLYGENIC EMBRYO SCREENING SERVICES.

POPULATION GENOMICS LITERATURE ON GENOME-WIDE ASSOCIATION STUDIES AND TRAIT HERITABILITY.

BIOETHICS JOURNALS ADDRESSING CONSUMER-DRIVEN SELECTION AND REPRODUCTIVE GENOMICS.

5.7 Beyond Selection, Germline Editing, In Vitro Gametogenesis, and the Threshold of Design

Up to this point, assisted reproduction has operated within a bounded logic. Technologies intervene to select among naturally occurring variation, not to create new variation outright. Even polygenic screening, for all its reach, remains fundamentally curatorial. It chooses between embryos that already exist.

The next class of technologies breaks that boundary.

Germline editing and in vitro gametogenesis do not ask which embryo should be transferred. They ask how embryos should be generated in the first place. This distinction marks the threshold between optimisation and design.

The scientific foundations for this shift are already established.

Germline Editing: From Error Correction to Alteration

Germline gene editing refers to the intentional modification of genetic material in gametes, embryos, or their precursor cells in a way that is heritable by future generations. The enabling technology most closely associated with this capability is CRISPR-Cas9, a genome-editing system adapted from bacterial immune mechanisms.

CRISPR's relevance to reproduction lies in its precision and accessibility. Earlier gene-editing methods were complex, expensive, and error-prone. CRISPR lowered technical barriers dramatically. It allowed targeted changes at specific genomic loci with a level of control previously unavailable.

The implications for human reproduction became clear almost immediately.

In 2018, He Jiankui announced the birth of genetically edited children in China, claiming to have altered the CCR5 gene to confer resistance to HIV. The scientific community responded with near-universal condemnation, not because germline editing was impossible, but because it had been done prematurely, secretly, and without adequate ethical or medical justification.

THE EPISODE IS OFTEN FRAMED AS AN ABERRATION. STRUCTURALLY, IT WAS A PREVIEW.

FROM A TECHNICAL STANDPOINT, GERMLINE EDITING OFFERS TWO POTENTIAL PATHWAYS. THE FIRST IS ERROR CORRECTION: REPAIRING PATHOGENIC MUTATIONS THAT WOULD OTHERWISE CAUSE SEVERE MONOGENIC DISEASE. THE SECOND IS TRAIT MODIFICATION: ALTERING GENES ASSOCIATED WITH NON-LETHAL TRAITS OR PROBABILISTIC RISKS.

THE FORMER IS FREQUENTLY PRESENTED AS ETHICALLY DEFENSIBLE. THE LATTER REMAINS HIGHLY CONTESTED.

HOWEVER, THE BOUNDARY BETWEEN CORRECTION AND ENHANCEMENT IS NOT FIXED. MANY CONDITIONS EXIST ON A SPECTRUM. REDUCING GENETIC RISK FOR CARDIOVASCULAR DISEASE, FOR EXAMPLE, MAY BE FRAMED AS PREVENTION RATHER THAN ENHANCEMENT. INCREASING RESISTANCE TO INFECTION MAY BE FRAMED AS HEALTH OPTIMISATION RATHER THAN ALTERATION.

ONCE GERMLINE EDITING IS TECHNICALLY FEASIBLE AND CLINICALLY SAFE, THE LIMITING FACTOR BECOMES GOVERNANCE RATHER THAN CAPABILITY.

AT PRESENT, MOST COUNTRIES PROHIBIT CLINICAL GERMLINE EDITING FOR REPRODUCTIVE PURPOSES. INTERNATIONAL SCIENTIFIC BODIES, INCLUDING THE WORLD HEALTH ORGANIZATION AND NATIONAL ACADEMIES OF SCIENCE, HAVE CALLED FOR MORATORIA OR STRICT REGULATION. THESE CONSTRAINTS ARE POLITICAL AND ETHICAL, NOT TECHNICAL.

Research continues regardless.

Laboratory experiments have demonstrated successful gene editing in human embryos under research protocols, typically without implantation. Animal models show heritable transmission of edited traits across generations. The knowledge base is expanding even as clinical application remains constrained.

From a fertility systems perspective, germline editing represents selection displaced upstream. Instead of choosing among embryos, the system alters the pool from which embryos arise. The logic of optimisation persists, but the mechanism changes.

This logic becomes even more pronounced when paired with a second technology.

In Vitro Gametogenesis: Rewriting the Source Material

In vitro gametogenesis, or IVG, refers to the generation of functional sperm or eggs from pluripotent stem cells rather than from mature gonads. In essence, IVG aims to decouple reproduction from the biological constraints of age, sex, and fertility status.

The scientific groundwork for IVG has advanced rapidly over the past two decades.

In animal models, researchers have successfully generated viable gametes from stem cells and produced healthy offspring. In mice, IVG has been used to create eggs from male cells and sperm from female cells, demonstrating the theoretical possibility of same-sex reproduction at the genetic level.

Human IVG remains experimental, but progress is tangible.

Research groups at institutions such as Kyoto University have reported the derivation of primordial germ cell-like cells from human induced pluripotent stem cells. These cells resemble early gamete precursors but have not yet been matured into fully functional sperm or eggs suitable for fertilisation.

Private sector involvement has accelerated development.

Companies such as Conception Biosciences and Gameto are pursuing IVG-related technologies with the explicit goal of extending reproductive lifespan and reducing age-related infertility. Their public communications emphasise therapeutic intent rather than enhancement, focusing on women with diminished ovarian reserve or premature ovarian failure.

Nevertheless, the implications are broader.

If gametes can be produced from stem cells, the constraints that currently define reproduction begin to dissolve. Age limits become flexible. Gamete quantity becomes scalable. Individuals without viable sperm or eggs gain new pathways to genetic parenthood. The concept of a finite ovarian reserve ceases to be absolute.

IVG also interacts directly with selection logic.

If large numbers of gametes can be generated in vitro, the number of embryos that can be created increases dramatically. Selection pressure intensifies. Screening becomes more granular. The reproductive funnel narrows further.

In this context, polygenic screening and germline editing become complementary rather than separate. IVG supplies volume. Screening supplies ranking. Editing supplies modification.

Together, they form a closed optimisation loop.

The Threshold of Design

What distinguishes this technological constellation from earlier assisted reproduction is not merely power, but intentionality.

IVF compensates for failure. Screening reduces risk. Germline editing and IVG enable proactive shaping of genetic outcomes. The reproductive process moves from reactive to anticipatory.

This shift challenges foundational assumptions.

Natural reproduction accepts randomness as given. Assisted reproduction already attenuates that randomness. Design-oriented technologies seek to replace it with control. The moral weight of this replacement does not lie solely in what is changed, but in what is no longer left to chance.

Importantly, none of these technologies require coercion to reshape reproduction. They align with existing incentives.

Patients already seek to minimise risk. Clinics already optimise outcomes. Markets already reward efficiency. Design-oriented reproduction extends these patterns rather than contradicting them.

The principal barrier is not demand, but legitimacy.

Public discomfort with germline editing often focuses on extreme scenarios: designer babies, enhanced intelligence, or engineered inequality. These scenarios distract from more likely early applications, which mirror past transitions. Disease avoidance precedes trait selection. Therapeutic framing precedes elective use. Exceptional cases normalise routine ones.

The trajectory is familiar.

From a systems perspective, the threshold of design is crossed not when editing becomes universal, but when it becomes ordinary enough to justify infrastructure. Research pipelines, regulatory frameworks, insurance coverage, and professional training all respond to perceived inevitability rather than ethical clarity.

IVG and germline editing remain constrained today, but they are no longer speculative in the scientific sense. They are speculative only in governance.

This distinction matters.

Once reproduction can be designed, even partially, the ethical centre of fertility medicine shifts again. The question is no longer how to overcome infertility, but what responsibility accompanies the power to choose not just embryos, but genomes.

The technologies that follow push this question further still.

If embryos and gametes can be generated and modified independently of the human body, then the body itself becomes optional in the reproductive process. Gestation, long assumed to be the final biological anchor, becomes the next site of intervention.

That step completes the arc.

References

World Health Organization. Human Genome Editing: Recommendations, 2021.

National Academies of Sciences. Reports on germline genome editing.

Nature and Science. Research articles on CRISPR-Cas9 and human embryo editing.

Human Reproduction Update. Reviews on in vitro gametogenesis progress and limitations.

Kyoto University research publications on primordial germ cell derivation.

Conception Biosciences. Public disclosures on IVG research direction.

Gameto Inc. Technical descriptions of egg cell engineering approaches.

Bioethics literature on germline intervention and reproductive design.

Chapter Six

Anxiety, Autonomy, and the Weight of Choice

When the Body Stops Being Assumed

I didn't notice the anxiety arrive. It didn't announce itself as fear. It crept in sideways, as a kind of restlessness that made it harder to sit still with the material I was reading.

Up to that point, everything had felt incremental. Dating platforms. Tracking. Diagnostics. Labs. Selection. Each step followed logically from the last. Even when the implications were unsettling, they were familiar in shape. Optimisation. Risk reduction. Efficiency. These were concepts I'd lived alongside for years without feeling personally threatened by them.

Something shifted when the body itself started to feel optional.

Not medically optional, but structurally optional. As if it were no longer the anchor of the process, just one possible interface among others. I caught myself rereading passages I'd already understood, not because they were unclear, but because I didn't like where they were leading.

I remembered books from university that I'd treated as theoretical at the time. Dense, angry texts that framed the body as a site of control and resistance. I'd read them with a kind of academic detachment, confident that the world they described was exaggerated, symbolic, safely abstract.

The arguments had felt loud and self-serious, more about ideology than reality.

What unsettled me now was how quiet the real version was.

There were no slogans in the material I was reading. No manifestos. No calls to arms. Just companies, protocols, roadmaps, and timelines. Everything framed as help. Everything couched in care. The horror, when it appeared, did so without drama.

It showed up in sentences that treated the body as a limitation to be engineered around.

I started noticing my own body differently. Not critically, but defensively. The way you become aware of something once you realise it might be taken for granted. I paid attention to textures. The heaviness of my limbs at the end of the day. The faint, persistent discomforts I'd always ignored because they were ordinary. Ordinary suddenly felt important.

Body autonomy is an easy phrase to use casually. It's harder to hold onto when autonomy dissolves not through force, but through irrelevance.

No one was threatening to take my body away. That wasn't the fear. The fear was that it would simply stop mattering to the process it had once defined.

I thought of friends who had done everything right. Education. Careers. Relationships carefully paced and responsibly delayed. People who assumed, as I had, that time would remain negotiable. Some of them had children eventually, after effort and expense that still surprised them. Others hadn't. Not because they didn't want to, but because wanting had quietly become insufficient.

Their stories had always sounded like cautionary tales when told retrospectively, but never like warnings while they were unfolding. No one stands in their twenties or thirties feeling reckless about waiting. Waiting feels sensible. It feels like maturity.

What no one tells you is that waiting doesn't announce when it turns into loss. It just changes the arithmetic.

As the technologies grew more radical on the page, my reactions became less analytical and more visceral. I wasn't worried about designer babies or science fiction extremes. I was worried about the narrowing of acceptable outcomes. About a future in which deviation from an optimised path feels irresponsible rather than human.

The word "horror" crossed my mind once, unexpectedly, and then stayed there.

Not horror as spectacle. Horror as quiet realisation. The kind that doesn't make you recoil, but makes you sit very still.

I understood, intellectually, that none of this was inevitable. Regulation exists. Ethics committees exist. Social limits exist. But I also understood how systems behave when incentives align. How quickly exceptional measures become routine when they solve real problems efficiently.

The most unsettling part was recognising how persuasive the logic was. I couldn't dismiss it as madness or cruelty. It was rational. It was compassionate, even. It was built to spare people pain.

And yet, somewhere along the way, it asked for a trade I wasn't sure I was willing to make.

Not a political trade. A biological one.

The body has always been inconvenient. It bleeds, it ages, it fails unpredictably. Technology has spent centuries trying to make those failures manageable. I've benefited from that more times than I can count. But reproduction sits at a different threshold. It isn't just something the body does. It's the point at which the body becomes generative rather than functional.

LOSING THAT CENTRALITY FELT LIKE LOSING SOMETHING OLDER THAN CHOICE.

I DIDN'T WANT TO STOP READING. I DIDN'T WANT TO LOOK AWAY. BUT I REALISED I NEEDED TO SLOW DOWN. TO LET THE WEIGHT OF WHAT WAS BEING PROPOSED REGISTER SOMEWHERE OTHER THAN MY INTELLECT.

THE SPECULATIVE CHAPTERS WERE COMING NEXT. ARTIFICIAL WOMBS. NON-HUMAN GESTATION. FUTURES WHERE THE BODY IS NO LONGER THE DEFAULT SITE OF BEGINNING. I COULD ALREADY SEE HOW NEATLY THEY FOLLOWED FROM EVERYTHING BEFORE.

WHAT FRIGHTENED ME WASN'T THAT THESE IDEAS EXISTED.

IT WAS THAT THEY MADE TOO MUCH SENSE.

BEFORE GOING ANY FURTHER, I NEEDED TO ACKNOWLEDGE THE FEAR PROPERLY. NOT TO RESOLVE IT. NOT TO DRAMATISE IT. JUST TO ADMIT THAT SOMEWHERE BETWEEN HELP AND HORROR, THERE IS A LINE THAT FEELS PERSONAL EVEN WHEN IT ISN'T.

THIS WAS WHERE I STOOD NOW. NOT OPPOSED. NOT CONVINCED. JUST AWARE THAT ONCE REPRODUCTION LEAVES THE BODY BEHIND, IT MAY NEVER FULLY COME BACK.

Chapter 7

Beyond the Body

Externalised Reproduction, Exogestation, and the Limits of Technology

7.1 What we are actually talking about

Defining ectogenesis, exogestation, and the boundaries of the real

The term "artificial womb" is one of the most misleading phrases currently circulating in public discourse around reproductive technology. It collapses multiple distinct scientific and clinical concepts into a single image, usually a glass tank containing a fully formed baby. That image is emotionally potent, but technically incoherent.

To understand what is actually being developed, and therefore what is investable, ethically defensible, and biologically plausible, it is necessary to separate three ideas that are often conflated.

Partial ectogenesis (exogestation)

The first and only near-term clinically realistic concept is partial ectogenesis, more accurately described as exogestation. This refers to the continuation of fetal development outside the human body after pregnancy has already begun but before the fetus is physiologically ready for neonatal life.

In practical terms, this means supporting a fetus that would otherwise be born extremely prematurely, typically between 22 and 26 weeks of gestation, in an environment that preserves fetal rather than neonatal physiology. The defining feature is that the fetus does not transition to air breathing, does not rely on immature lungs, and does not undergo the abrupt circulatory changes that normally occur at birth.

This is not an attempt to replace pregnancy. It is an attempt to avoid forcing a biological transition too early.

The medical motivation is clear and widely accepted. Extreme prematurity is one of the leading causes of infant mortality and long-term disability globally. Survivors often face chronic lung disease, neurodevelopmental impairment, gastrointestinal complications, and lifelong health burdens. Neonatal intensive care units are highly advanced, but they are fundamentally designed to support newborns, not fetuses.

Exogestation proposes a different strategy: instead of pushing the fetus forward prematurely, hold it back physiologically while time catches up.

This framing is crucial. Nearly all serious academic and clinical teams avoid the term "artificial womb" in formal literature precisely because it invites misunderstanding.

They describe their work as artificial placenta systems, extrauterine fetal support, or perinatal life support. The language signals continuity with pregnancy, not replacement of it.

From a regulatory and ethical standpoint, this distinction is not cosmetic. It defines whether a technology is seen as an extension of neonatal care or a redefinition of reproduction itself.

Complete ectogenesis

The second concept is complete ectogenesis, meaning the initiation and completion of gestation entirely outside the human body, from fertilisation to delivery. This would require not only sustaining fetal development, but replicating implantation, placentation, immune tolerance, endocrine signalling, and developmental timing across months.

At present, this is not a credible near-term engineering target.

The barriers are not limited to hardware. Human pregnancy is a deeply interactive biological process involving dynamic signalling between maternal tissues and the developing fetus. The placenta is not a passive filter. It is an endocrine organ, an immune modulator, a nutrient sensor, and a regulatory interface. Replicating its full functionality outside the body would require solving multiple unsolved problems in developmental biology simultaneously.

This is why credible academic literature on ectogenesis overwhelmingly focuses on partial systems. When complete ectogenesis appears, it is typically in speculative ethics papers, philosophical discussions, or media coverage extrapolating far beyond present capability.

For the purposes of a serious technology and investment analysis, complete ectogenesis should be treated as conceptual background, not an active development pipeline.

Neonatal intensive care is not ectogenesis

The third idea that requires separation is conventional neonatal intensive care. NICUs already support extremely premature infants with ventilation, intravenous nutrition, and incubators. However, this is not ectogenesis. It is an emergency adaptation that forces a fetus to behave like a neonate before it is ready.

The key difference lies in physiology.

Neonatal care assumes air breathing, pulmonary circulation, and digestive function. Exogestation systems explicitly aim to avoid these transitions. They maintain fetal circulation patterns, rely on extracorporeal gas exchange rather than lungs, and keep the fetus submerged in a fluid environment analogous to amniotic fluid.

This distinction explains why artificial placenta research emerged not from neonatal incubator design, but from cardiovascular engineering, extracorporeal life support, and fetal surgery research.

Why definitions matter for investment and ethics

This definitional clarity matters because it draws a boundary between medical necessity and speculative redesign of reproduction.

Partial ectogenesis addresses a clearly defined clinical problem with measurable endpoints: survival rates, neurodevelopmental outcomes, and complication reduction in extremely premature infants. It aligns with existing hospital infrastructure, reimbursement frameworks, and regulatory pathways.

Complete ectogenesis, by contrast, would redefine reproduction itself and trigger legal, ethical, and cultural questions that no current health system is prepared to resolve.

Investors, policymakers, and readers need to understand that the field is not marching uniformly toward a single endpoint. It is expanding laterally, building tools that extend gestation slightly longer under controlled conditions, not vertically toward total externalisation of pregnancy.

This lateral expansion is why the field attracts serious institutional funding rather than fringe capital. It is framed as neonatal care evolution, not reproductive revolution.

The strategic wedge

Every major programme in this space uses the same wedge argument:

If you can safely maintain fetal physiology for even two to four additional weeks in cases of extreme prematurity, the impact on survival and lifelong outcomes is disproportionate.

This is not a moonshot claim. It is grounded in epidemiology. Outcomes improve sharply with each additional week in utero across the viability threshold. Exogestation systems aim to capture that gradient.

Understanding this wedge is essential. It explains why the field is conservative in language, incremental in goals, and cautious in public framing, even as the underlying implications are profound.

This chapter, therefore, is not about science fiction. It is about a narrow, technically demanding, ethically heavy domain where small extensions of biology create large downstream effects.

That is the terrain we are now entering.

References

Defining ectogenesis, exogestation, and the boundaries of the real

Partridge, E.A., Davey, M.G., Hornick, M.A., McGovern, P.E., Mejaddam, A.Y., Vrecenak, J.D., Mesas-Burgos, C., Olive, A., Caskey, R.C., Weiland, T.R., Han, J., Schupper, A.J., Connelly, J.T., Dysart, K.C., Rychik, J., Hedrick, H.L., Flake, A.W. and Hirschl, R.B. (2017) 'An extra-uterine system to physiologically support the extreme premature lamb', Nature Communications, 8, 15112.

Flake, A.W. and Rychik, J. (2021) 'Artificial placenta technology for the treatment of extreme prematurity', Seminars in Fetal and Neonatal Medicine, 26(2), 101215.

Romanis, E.C. (2018) 'Artificial womb technology and the frontiers of human reproduction: Conceptual differences and potential implications', Journal of Medical Ethics, 44(11), pp. 751–755.

Romanis, E.C. (2020) 'Challenging the binary: Artificial wombs and the reshaping of pregnancy', Medical Law Review, 28(1), pp. 1–30.

Hornick, M.A., Davey, M.G., Partridge, E.A., McGovern, P.E., Mejaddam, A.Y., Vrecenak, J.D., Hedrick, H.L., Rychik, J., Flake, A.W. and Hirschl, R.B. (2019) 'An experimental artificial placenta supports normal fetal development in premature lambs', Journal of Thoracic and Cardiovascular Surgery, 155(2), pp. 778–788.

Miura, Y., Matsuda, T. and Funamoto, K. (2019) 'Current challenges and perspectives of artificial placenta development', Artificial Organs, 43(5), pp. 447–458.

De Bie, F.R., Davey, M.G., Vrecenak, J.D., McGovern, P.E., Hornick, M.A., Partridge, E.A., Flake, A.W. and Hirschl, R.B. (2020) 'The artificial placenta: Where are we now?', Pediatric Research, 87, pp. 775–782.

Kozlov, M. (2023) 'Human trials of artificial wombs could start soon', Nature, 613, pp. 222–224.

Perinatal Life Support Consortium (2020) Perinatal Life Support: An incubator with intact physiology for extremely premature infants, Horizon 2020 Project Overview, European Commission (CORDIS).

World Health Organization (2015) WHO recommendations on interventions to improve preterm birth outcomes, WHO Guidelines.

7.2 The Clinical and Demographic Case for Exogestation

Why health systems are willing to fund something this hard

Exogestation exists because extreme prematurity remains one of the most stubborn, expensive, and morally charged problems in modern medicine. It sits at the intersection of biology, probability, and time, and it resists simple solutions. Neonatal care has improved dramatically over the past half century, but it has not altered the fundamental constraint that fetal organs mature on a schedule that technology cannot easily accelerate.

The demographic reality is stark.

Globally, an estimated 15 million babies are born preterm each year, defined as birth before 37 weeks of gestation. Of these, roughly 1 to 2 million are born extremely preterm, typically before 28 weeks. Survival improves rapidly with each additional week of gestation, but below approximately 26 weeks, outcomes become highly variable and deeply dependent on local resources, protocols, and luck.

This is not evenly distributed.

Rates of prematurity are higher in low and middle income countries, but the ethical pressure around exogestation is driven primarily by high income health systems, where survival is already possible but often accompanied by severe long-term morbidity. In these systems, the question is no longer whether survival can be achieved at the edge of viability, but what kind of life that survival produces.

Extreme prematurity is associated with a predictable cluster of complications. These include bronchopulmonary dysplasia due to immature lungs exposed to positive pressure ventilation, intraventricular haemorrhage and white matter injury in the brain, necrotising enterocolitis of the gut, retinopathy of prematurity, and long-term neurodevelopmental impairment. These outcomes are not rare edge cases. They are the statistical norm at the earliest gestational ages.

From a systems perspective, this creates a compound cost problem.

First, there is the immediate cost of neonatal intensive care. NICU stays for extremely preterm infants often last months and require highly specialised staff, equipment, and round-the-clock monitoring. Second, there are the long-term costs. Children who survive extreme prematurity may require ongoing medical care, educational support, and social services for decades. These costs are borne not only by families, but by health systems and states.

Health economics studies consistently show that even modest reductions in severe disability rates among extremely preterm survivors can justify substantial upfront investment. This is one of the reasons why exogestation attracts public funding rather than being left solely to speculative private capital.

The clinical logic behind exogestation is not radical. It is conservative.

Neonatal intensive care forces a fetus to adapt prematurely to an environment it is not developmentally prepared for. The lungs are asked to exchange gas before surfactant production is adequate. The gut is asked to digest before it has matured. The brain is exposed to pressure fluctuations, oxygen variability, and inflammatory insults at a vulnerable stage.

Exogestation attempts to avoid these insults by maintaining fetal physiology rather than accelerating neonatal adaptation.

The key insight is that survival is not the only metric that matters. The quality of survival is equally important. A technology that modestly improves survival but significantly reduces rates of severe disability can be more valuable, ethically and economically, than one that simply pushes viability earlier.

This is why most serious research programmes frame their goal not as "earlier viability," but as a bridge across the most dangerous developmental window.

From a demographic standpoint, this bridge is narrow but powerful. The number of infants who would qualify for exogestation is relatively small compared to total births. This makes the problem technically demanding but economically tractable. Health systems are accustomed to funding rare, high-complexity interventions when the burden of disease is severe and alternatives are limited.

There is also a legal and professional dimension.

In many high income countries, clinicians already face intense scrutiny around decisions at the edge of viability. Parents are asked to consent to aggressive interventions with uncertain outcomes under conditions of acute stress. Litigation risk, ethical review, and public attention all converge in this space. Technologies that reduce uncertainty or improve outcomes, even incrementally, are attractive to institutions that must operate under these pressures.

Importantly, exogestation does not remove hard decisions. It reshapes them.

THE DECISION POINT SHIFTS FROM "DO WE RESUSCITATE" TO "DO WE INITIATE ARTIFICIAL PLACENTAL SUPPORT." THIS REFRAMING MATTERS BECAUSE IT ALIGNS INTERVENTION WITH AN INTUITIVE BIOLOGICAL NARRATIVE: WE ARE CONTINUING PREGNANCY, NOT REPLACING IT. THAT NARRATIVE IS EASIER FOR CLINICIANS AND FAMILIES TO ACCEPT, AND EASIER FOR REGULATORS TO JUSTIFY.

DEMOGRAPHICS ALSO INTERACT WITH BROADER SOCIAL TRENDS.

IN MANY COUNTRIES, MATERNAL AGE AT FIRST BIRTH HAS RISEN STEADILY. ADVANCED MATERNAL AGE IS ASSOCIATED WITH HIGHER RATES OF COMPLICATIONS, INCLUDING PRETERM BIRTH. WHILE EXOGESTATION DOES NOT ADDRESS THE CAUSES OF PREMATURITY, IT AMPLIFIES THE CONSEQUENCES OF DEMOGRAPHIC DELAY BY INCREASING THE NUMBER OF INFANTS WHO MIGHT BENEFIT FROM IMPROVED SUPPORT AT EARLY GESTATIONS.

THIS CREATES A SUBTLE FEEDBACK LOOP.

AS SOCIETIES TOLERATE LATER PARENTHOOD, HEALTH SYSTEMS FACE HIGHER PREMATURITY BURDENS. AS THOSE BURDENS RISE, PRESSURE GROWS TO DEVELOP TECHNOLOGIES THAT MITIGATE DOWNSTREAM HARM. EXOGESTATION EMERGES NOT AS A LIFESTYLE TECHNOLOGY, BUT AS A DAMAGE CONTROL SYSTEM FOR MODERN REPRODUCTIVE TIMING.

FROM AN INVESTMENT PERSPECTIVE, THIS DEMOGRAPHIC AND CLINICAL FRAMING IS ESSENTIAL.

The addressable market is not "all pregnancies." It is a tightly defined subset of high-acuity cases with high willingness to pay, strong public funding support, and clear outcome metrics. This is the same profile that has historically supported successful adoption of technologies such as extracorporeal membrane oxygenation, advanced neonatal ventilation systems, and paediatric cardiac interventions.

The moral force of the problem should not be understated.

Extreme prematurity forces societies to confront uncomfortable questions about how much intervention is too much, and how to weigh survival against suffering. Exogestation does not resolve these questions. It offers a new way to approach them, grounded in the idea that time itself is the missing therapy.

By buying time without forcing premature adaptation, exogestation reframes the problem from one of heroic rescue to one of physiological patience.

That reframing is why the field exists at all.

It is also why, once introduced, it will be very difficult to abandon.

References

The clinical and demographic case for exogestation

Blencowe, H., Cousens, S., Oestergaard, M.Z., Chou, D., Moller, A.B., Narwal, R., Adler, A., Vera Garcia, C., Rohde, S., Say, L. and Lawn, J.E. (2012) 'National, regional, and worldwide estimates of preterm birth rates in the year 2010', The Lancet, 379(9832), pp. 2162–2172.

World Health Organization (2023) Preterm birth. Available via WHO fact sheets and epidemiological summaries.

Saigal, S. and Doyle, L.W. (2008) 'An overview of mortality and sequelae of preterm birth from infancy to adulthood', The Lancet, 371(9608), pp. 261–269.

Stoll, B.J., Hansen, N.I., Bell, E.F., Walsh, M.C., Carlo, W.A., Shankaran, S., Laptook, A.R., Sánchez, P.J., Van Meurs, K.P., Wyckoff, M., Das, A. and Higgins, R.D. (2015) 'Trends in care practices, morbidity, and mortality of extremely preterm neonates, 1993–2012', JAMA, 314(10), pp. 1039–1051.

Rysavy, M.A., Li, L., Bell, E.F., Das, A., Hintz, S.R., Stoll, B.J., Vohr, B.R., Carlo, W.A., Shankaran, S., Walsh, M.C., Laptook, A.R., Sánchez, P.J., Van Meurs, K.P., Wyckoff, M., Higgins, R.D. and Eunice Kennedy Shriver National Institute of Child Health and

Human Development Neonatal Research Network (2015) 'Between-hospital variation in treatment and outcomes in extremely preterm infants', New England Journal of Medicine, 372(19), pp. 1801–1811.

Moore, T., Hennessy, E.M., Myles, J., Johnson, S.J., Draper, E.S., Costeloe, K.L. and Marlow, N. (2012) 'Neurological and developmental outcome in extremely preterm children born in England in 1995 and 2006', BMJ, 345, e7961.

Petrou, S., Eddama, O. and Mangham, L. (2011) 'A structured review of the recent literature on the economic consequences of preterm birth', Archives of Disease in Childhood: Fetal and Neonatal Edition, 96(3), pp. F225–F232.

Partridge, E.A., Davey, M.G., Hornick, M.A., McGovern, P.E., Mejaddam, A.Y., Vrecenak, J.D., Mesas-Burgos, C., Olive, A., Caskey, R.C., Weiland, T.R., Han, J., Schupper, A.J., Connelly, J.T., Dysart, K.C., Rychik, J., Hedrick, H.L., Flake, A.W. and Hirschl, R.B. (2017) 'An extra-uterine system to physiologically support the extreme premature lamb', Nature Communications, 8, 15112.

Kozlov, M. (2023) 'Human trials of artificial wombs could start soon', Nature, 613, pp. 222–224.

Eichler, F., Huppi, P.S. and Hagmann, C. (2017) 'Preterm birth and brain development', The Lancet Neurology, 16(2), pp. 147–159.

European Commission (2020) Perinatal Life Support (PLS): Grant Agreement No. 863087, Horizon 2020 Programme documentation (CORDIS).

7.3 The Engineering Stack

Artificial placentas, circulation physics, materials science, and why this is a systems problem

If exogestation were merely a question of keeping a fetus warm and supplied with oxygen, it would already exist. What makes it hard is that pregnancy is not a single function but a stack of interdependent biological systems that must remain synchronised. Artificial womb research fails when any one layer is treated in isolation.

Successful platforms therefore converge on the same conclusion: this is not a device, it is an integrated life-support architecture.

Preserving fetal physiology rather than simulating neonatal life

The defining engineering principle of exogestation systems is the preservation of fetal circulation. In utero, the fetus does not rely on the lungs for gas exchange. Oxygenated blood arrives via the placenta, bypassing pulmonary circulation through the ductus arteriosus and foramen ovale. Birth forcibly rewires this system.

Neonatal intensive care accepts this forced transition and attempts to support it. Artificial placenta systems do the opposite. They attempt to avoid the transition altogether.

This choice drives every downstream engineering decision. If fetal physiology is to be preserved, then oxygen delivery must occur extracorporeally, circulation must remain low resistance, and pressure gradients must not exceed what the fetal heart can tolerate.

The fetal heart is not a pump designed for extracorporeal circuits. Its output is limited, its pressure tolerance is low, and its response to afterload is narrow. Any artificial system that imposes excessive resistance risks cardiac failure. This is why nearly all credible designs aim for pumpless arteriovenous circulation, driven entirely by the fetal heart.

This single requirement eliminates large classes of conventional life-support hardware from consideration.

Artificial placenta oxygenation: gas exchange without shear

At the core of exogestation platforms is the artificial placenta, an extracorporeal oxygenation unit that must deliver sufficient gas exchange at extremely low resistance. Conventional extracorporeal membrane oxygenation systems are unsuitable. They are designed for adults or neonates with higher pressures and tolerate higher shear forces.

Artificial placenta oxygenators therefore require:

- Ultra-low resistance flow paths
- High surface area gas exchange membranes
- Minimal turbulence and shear stress
- Materials that do not trigger coagulation or inflammation

Most designs rely on hollow-fiber membrane bundles, but scaled and configured differently from standard ECMO. Fiber diameter, packing density, surface coatings, and flow geometry become critical variables.

Experimental papers consistently report that small increases in circuit resistance produce disproportionate physiological instability in fetal models. This is why engineering papers in the field read more like cardiovascular physics than neonatal medicine.

From an investment perspective, this is a key insight. The value is not only in the final system, but in component technologies: membranes, coatings, connectors, and microfluidic geometries that can deliver low-resistance exchange reliably and reproducibly.

This is where suppliers of advanced polymers, surface treatments, and precision manufacturing quietly matter more than brand-name "artificial womb" startups.

Blood compatibility, clotting, and the anticoagulation dilemma

Blood does not like artificial surfaces. Any extracorporeal circuit risks clot formation, platelet activation, and inflammatory cascades. In adults, anticoagulation can be aggressive. In fetuses, anticoagulation margins are extremely narrow.

This creates a fundamental design tension.

On one side is the need to prevent clotting in the circuit. On the other is the need to avoid haemorrhage, particularly intracranial bleeding, which is already a major risk in extreme prematurity.

Engineering responses to this tension focus heavily on surface science rather than pharmacology. Heparin-bonded surfaces, endothelial-mimicking coatings, and ultra-smooth flow paths are all actively explored. Reducing contact activation is preferable to suppressing coagulation systemically.

This emphasis explains why artificial placenta research overlaps strongly with vascular graft research, cardiac assist device development, and haemocompatible materials science.

From a commercial standpoint, any surface technology that demonstrably reduces thrombogenicity without systemic anticoagulation has value far beyond exogestation. This is a classic platform spillover opportunity.

Umbilical vascular access: the non-negotiable failure point

If oxygenation is the heart of the system, umbilical cannulation is its Achilles' heel.

The umbilical vessels are small, fragile, and not designed for prolonged external access. Cannulation must be secure, stable,

AND MINIMALLY TRAUMAT
IC. LOSS OF ACCESS IS CATASTROPHIC. INFECTION RISK IS EXISTENTIAL. MECHANICAL DISLODGEMENT CANNOT BE TOLERATED.

MOST EXPERIMENTAL SYSTEMS RELY ON CAREFULLY DEVELOPED CANNULATION TECHNIQUES PERFORMED IMMEDIATELY AFTER DELIVERY, BEFORE VESSEL CONSTRICTION. THIS REQUIRES SPECIALISED SURGICAL SKILL AND PROTOCOL DISCIPLINE.

FROM A SYSTEMS PERSPECTIVE, THIS IS ONE OF THE STRONGEST ARGUMENTS AGAINST CASUAL DEPLOYMENT. EXOGESTATION IS NOT SOMETHING THAT CAN BE IMPROVISED. IT REQUIRES TRAINED TEAMS, DEFINED WORKFLOWS, AND INSTITUTIONAL COMMITMENT.

IT ALSO HIGHLIGHTS ANOTHER INVESTMENT VECTOR: PROCEDURAL TOOLING AND TRAINING ECOSYSTEMS. DEVICES THAT MAKE UMBILICAL ACCESS SAFER, FASTER, AND MORE REPRODUCIBLE ARE NOT OPTIONAL EXTRAS. THEY ARE GATING TECHNOLOGIES.

THE FLUID ENVIRONMENT: AMNIOTIC SIMULATION IS NOT COSMETIC

KEEPING THE FETUS SUBMERGED IN A FLUID ENVIRONMENT IS OFTEN MISINTERPRETED AS AESTHETIC OR SYMBOLIC. IN REALITY, IT IS MECHANICALLY AND BIOLOGICALLY NECESSARY.

Amniotic fluid supports musculoskeletal development, prevents lung collapse, stabilises temperature, and reduces sensory overload. Air exposure introduces evaporative heat loss, pressure on immature tissues, and mechanical stress that the fetus is not prepared to handle.

Artificial systems therefore maintain a sealed, temperature-controlled, sterile fluid chamber with continuous fluid exchange. The fluid itself must be carefully formulated to avoid toxicity, electrolyte imbalance, and microbial growth.

This again expands the engineering stack. Sterile fluid management, filtration, monitoring, and replacement are not trivial at multi-day or multi-week timescales.

Failures here are subtle. Infection risk rises quietly. Electrolytes drift. Metabolic byproducts accumulate. This is why serious programmes emphasise closed-loop monitoring rather than static environments.

Monitoring, modelling, and closed-loop control

No artificial placenta system can be clinically viable without continuous monitoring and control. The fetus cannot report distress. The system must infer it.

Key parameters include blood flow, oxygenation, pressure gradients, heart rate variability, metabolic markers, and fluid composition. Sensors must be non-invasive where possible and reliable over long durations.

The emerging consensus is that exogestation platforms will require model-based control systems, not simple alarms. Physiological models help distinguish between transient fluctuation and true deterioration. This is particularly important because fetal physiology naturally differs from neonatal norms.

This is where software becomes central rather than ancillary.

Projects such as the EU-funded Perinatal Life Support programme explicitly frame modelling and decision support as core components, not add-ons. The aim is not automation for its own sake, but reduction of cognitive burden on clinical teams operating under extreme stress.

From a TechEyeSpy perspective, this is the moment where exogestation intersects with medical AI, digital twins, and regulatory-grade clinical software. These tools must be interpretable, auditable, and conservative. Flashy optimisation is a liability here.

Why this remains a systems problem, not a gadget

Each layer of the engineering stack interacts with the others. Reducing oxygenator resistance may alter flow dynamics. Changing surface coatings may affect infection risk. Adjusting fluid composition may influence cardiovascular stability.

This interdependence is why progress in the field appears slow to outsiders. Breakthroughs are incremental because they must be compatible with the entire system.

It also explains why translation from animal models to humans is cautious. Lamb models approximate fetal physiology, but scaling, anatomy, and gestational timing differ. Systems must be revalidated at every step.

For investors, this complexity is both risk and moat.

High barriers to entry protect serious teams and well-funded programmes. The downside is long timelines and regulatory drag. The upside is that partial success creates defensible platforms rather than commoditised products.

Exogestation engineering does not reward shortcuts.

It rewards patience, integration, and respect for biological constraint. Those traits are unfashionable in technology cycles, but they are precisely why this field attracts institutional money rather than hype capital.

References

Artificial placentas, circulation physics, and systems engineering

Partridge, E.A., Davey, M.G., Hornick, M.A., McGovern, P.E., Mejaddam, A.Y., Vrecenak, J.D., Mesas-Burgos, C., Olive, A., Caskey, R.C., Weiland, T.R., Han, J., Schupper, A.J., Connelly, J.T., Dysart, K.C., Rychik, J., Hedrick, H.L., Flake, A.W. and Hirschl, R.B. (2017) 'An extra-uterine system to physiologically support the extreme premature lamb', Nature Communications, 8, 15112.

Hornick, M.A., Davey, M.G., Partridge, E.A., McGovern, P.E., Mejaddam, A.Y., Vrecenak, J.D., Hedrick, H.L., Rychik, J., Flake, A.W. and Hirschl, R.B. (2019) 'An experimental artificial placenta supports normal fetal development in premature lambs', Journal of Thoracic and Cardiovascular Surgery, 155(2), pp. 778–788.

Flake, A.W., Villa, R.L., Rychik, J. and Hirschl, R.B. (2020) 'Artificial placenta: A next-generation neonatal life support system', Seminars in Fetal and Neonatal Medicine, 25(5), 101108.

Miura, Y., Matsuda, T. and Funamoto, K. (2019) 'Current challenges and perspectives of artificial placenta development', Artificial Organs, 43(5), pp. 447–458.

De Bie, F.R., Davey, M.G., Vrecenak, J.D., McGovern, P.E., Hornick, M.A., Partridge, E.A., Flake, A.W. and Hirschl, R.B. (2020) 'The artificial placenta: Where are we now?', Pediatric Research, 87, pp. 775–782.

Cao, H., McGovern, P.E., Davey, M.G., Hornick, M.A., Partridge, E.A., Flake, A.W. and Hirschl, R.B. (2023) 'Impact of circuit resistance on fetal hemodynamics in an artificial placenta system', Prenatal Diagnosis, 43(1), pp. 85–94.

Eixarch, E., Gratacós, E., Puig-Corveró, E., et al. (2023) 'Artificial placenta for extremely premature infants: Design and preclinical evaluation', Artificial Organs, 47(6), pp. 1034–1045.

Makdisi, G. and Wang, I.W. (2015) 'Extra corporeal membrane oxygenation (ECMO): Review of a lifesaving technology', Journal of Thoracic Disease, 7(7), pp. E166–E176.
(Used for comparative ECMO limitations versus artificial placenta requirements.)

Badylak, S.F. and Gilbert, T.W. (2008) 'Immune response to biologic scaffold materials', Seminars in Immunology, 20(2), pp. 109–116.
(Relevant to haemocompatible and endothelial-mimicking surface design.)

Ratner, B.D. (2019) 'The catastrophe revisited: Blood compatibility in the 21st century', Biomaterials, 232, 119704.

European Commission (2020) Perinatal Life Support (PLS): Integration of enabling technologies, Horizon 2020 Project No. 863087, CORDIS documentation.

Funamoto, K., Hayase, T. and Kato, M. (2012) 'Numerical analysis of blood flow in artificial placenta oxygenators', Journal of Biomechanical Science and Engineering, 7(2), pp. 153–165.

7.4 Who Is Actually Building This

Universities, hospitals, startups, and the translational pipeline from lab to clinic

Exogestation is not being built by lone inventors or speculative futurists. It is emerging from a tight triangle of actors that must cooperate for the field to advance at all: academic laboratories, paediatric hospitals, and translational startups. Each plays a distinct role, and misunderstanding those roles leads to exaggerated expectations about timelines and capabilities.

The most important structural fact is this: no artificial placenta system will ever be "launched" like a consumer product. These technologies move only when research credibility, clinical legitimacy, and regulatory plausibility advance together.

Academic centres as system architects

The intellectual centre of gravity for exogestation remains within academic medicine, not industry. This is not accidental. Artificial placenta systems demand long-duration animal studies, multidisciplinary collaboration, and tolerance for failure. These are poorly aligned with venture timelines

BUT WELL SUITED TO UNIVERSITY HOSPITALS.

THE MOST CITED LINEAGE ORIGINATES AT THE CHILDREN'S HOSPITAL OF PHILADELPHIA, WHERE TEAMS LED BY FETAL SURGEONS AND BIOENGINEERS DEVELOPED THE EXTEND SYSTEM. THEIR WORK COMBINED FETAL SURGERY EXPERTISE, CARDIOVASCULAR ENGINEERING, AND NEONATAL MEDICINE INTO A SINGLE PLATFORM. CRUCIALLY, CHOP'S ROLE WAS NOT JUST TECHNICAL. IT PROVIDED ETHICAL OVERSIGHT, INSTITUTIONAL REVIEW INFRASTRUCTURE, AND CREDIBILITY WITH REGULATORS.

ACADEMIC GROUPS IN EUROPE FOLLOW A SIMILAR MODEL. THE EINDHOVEN UNIVERSITY OF TECHNOLOGY COORDINATES THE PERINATAL LIFE SUPPORT PROGRAMME, A HORIZON-FUNDED CONSORTIUM THAT EXPLICITLY INTEGRATES ENGINEERING DEPARTMENTS, MEDICAL CENTRES, AND INDUSTRIAL PARTNERS. THIS STRUCTURE REFLECTS AN UNDERSTANDING THAT NO SINGLE DISCIPLINE CAN SOLVE THE PROBLEM ALONE.

OTHER EUROPEAN CENTRES, INCLUDING TEAMS ASSOCIATED WITH BCNATAL, HAVE FOCUSED ON ARTIFICIAL PLACENTA SUBSYSTEMS AND PRECLINICAL VALIDATION, PARTICULARLY IN LARGE-ANIMAL MODELS. THEIR WORK REINFORCES THE POINT THAT PROGRESS IS DISTRIBUTED RATHER THAN CENTRALISED.

THE ACADEMIC ROLE IS TO DE-RISK THE SCIENCE. UNIVERSITIES ESTABLISH PROOF OF PHYSIOLOGICAL PLAUSIBILITY, IDENTIFY FAILURE MODES, AND PUBLISH OPENLY. THEY ARE NOT OPTIMISED FOR MANUFACTURING, REGULATORY SUBMISSIONS, OR COMMERCIAL SCALING. THAT HANDOFF REQUIRES A DIFFERENT VEHICLE.

Hospitals as gatekeepers of reality

Paediatric hospitals occupy a uniquely powerful position in this ecosystem. They are where experimental systems confront real constraints: staffing, infection control, consent, liability, and workflow integration. Without hospital buy-in, exogestation remains a laboratory curiosity.

Hospitals also define the acceptable use case. It is not investors or ethicists who decide where these systems will first be deployed. It is neonatal teams facing extreme prematurity cases at 3 a.m., weighing risks under pressure.

This is why early translation efforts are anchored to major children's hospitals rather than private clinics. Neonatal intensive care units already manage extracorporeal support technologies such as ECMO. They understand the operational burden and the stakes.

Hospitals also shape regulation indirectly. Regulators do not approve devices in a vacuum. They evaluate them in the context of proposed clinical use. A system framed as an extension of neonatal life support is reviewed differently from one framed as a replacement for pregnancy.

This framing discipline originates at the hospital level.

Translational startups: narrow missions, heavy regulation

The role of startups in this space is often misunderstood. These are not moonshot companies promising to "end pregnancy." They are translation vehicles designed to carry a narrow, highly constrained technology across the regulatory valley of death.

A clear example is Vitara Biomedical, which emerged directly from the CHOP lineage. Vitara's stated aim is not complete ectogenesis, but the development of a regulated artificial placenta system capable of supporting extremely premature infants for a limited duration. Public communications emphasise incremental clinical milestones, not speculative futures.

This is deliberate. Venture capital in this space is structured around risk containment, not disruption narratives. Investors accept long timelines only when the endpoint is concrete: first-in-human trials, defined indications, and hospital procurement pathways.

European efforts show similar restraint. Startups such as Juno Perinatal Healthcare and AquaWomb have emerged from university consortia rather than independent invention. Their messaging consistently avoids claims of full gestation replacement, focusing instead on improving outcomes for extremely preterm infants.

This caution is not just ethical. It is strategic. Overstating capability invites regulatory resistance and public backlash, both of which can stall a field for years.

Funding sources reveal intent

Following the money clarifies motivations.

The presence of public funding, particularly through EU Horizon programmes and national research councils, signals that exogestation is being treated as a public health intervention rather than a private lifestyle technology. These funding bodies impose milestone reporting, ethical review, and cross-disciplinary collaboration requirements that constrain scope creep.

Private capital enters later, once proof-of-concept exists and regulatory pathways are visible. Even then, funding rounds tend to be smaller and more staged than in consumer biotech. This reflects the reality that exogestation systems will likely be reimbursed through health systems rather than sold directly.

Strategic partners also matter. Component suppliers in oxygenation, medical polymers, sensors, and clinical software often participate quietly through joint development agreements rather than headline acquisitions. Their interest lies in platform adjacency. A coating or membrane developed for artificial placenta use may later migrate into ECMO, dialysis, or vascular devices.

Why this pipeline is slow by design

From the outside, progress appears glacial. From inside the system, it is deliberate.

Each step in the pipeline imposes new constraints. Animal model success must translate to human anatomy. Human feasibility studies must demonstrate safety under worst-case conditions. Regulatory approval requires not only efficacy, but failure containment.

This slow pace is not a flaw. It is the price of operating in a domain where mistakes are irreversible and public trust is fragile.

Importantly, this also creates a natural moat. Few actors are willing or able to sustain decade-long development cycles with heavy ethical scrutiny and limited upside narratives. Those who do gain defensible positions not through speed, but through legitimacy.

The emerging pattern

Across regions, a consistent pattern emerges:

Academic institutions develop and validate systems

Hospitals define acceptable use and enforce discipline

Startups translate narrowly defined capabilities

Public funding stabilises early risk

Private capital enters cautiously at inflection points

This pattern explains why the field feels conservative despite its implications. It is not being driven by ideology or futurism. It is being pulled forward by unresolved clinical need.

For the reader, this matters because it reframes expectation. Exogestation is not a speculative endpoint waiting for permission. It is a constrained engineering response to a real problem, advancing under the weight of medical, legal, and ethical gravity.

The next section extends this logic.

Once systems exist, even in limited form, the question is no longer who is building them, but where

INVESTMENT AND CAPABILITY ACCUMULATE DOWNSTREAM: COMPONENTS, SOFTWARE, TRAINING, AND ADJACENT REPRODUCTIVE TECHNOLOGIES THAT BENEFIT FROM THE SAME SHIFT TOWARD EXTERNALISED BIOLOGICAL SUPPORT.

References

UNIVERSITIES, HOSPITALS, STARTUPS, AND THE TRANSLATIONAL PIPELINE

PARTRIDGE, E.A., DAVEY, M.G., HORNICK, M.A., MCGOVERN, P.E., MEJADDAM, A.Y., VRECENAK, J.D., HEDRICK, H.L., RYCHIK, J., FLAKE, A.W. AND HIRSCHL, R.B. (2017) 'AN EXTRA-UTERINE SYSTEM TO PHYSIOLOGICALLY SUPPORT THE EXTREME PREMATURE LAMB', NATURE COMMUNICATIONS, 8, 15112.
(FOUNDATIONAL CHOP WORK ESTABLISHING THE EXTEND SYSTEM AND INSTITUTIONAL MODEL.)

FLAKE, A.W. AND RYCHIK, J. (2021) 'ARTIFICIAL PLACENTA TECHNOLOGY FOR THE TREATMENT OF EXTREME PREMATURITY', SEMINARS IN FETAL AND NEONATAL MEDICINE, 26(2), 101215.
(EXPLICIT DISCUSSION OF TRANSLATIONAL CHALLENGES, HOSPITAL INTEGRATION, AND REGULATORY FRAMING.)

Hornick, M.A. et al. (2019) 'An experimental artificial placenta supports normal fetal development in premature lambs', Journal of Thoracic and Cardiovascular Surgery, 155(2), pp. 778–788.
(Details long-duration animal support and institutional infrastructure requirements.)

Kozlov, M. (2023) 'Human trials of artificial wombs could start soon', Nature, 613, pp. 222–224.
(Reports on translational readiness, hospital involvement, and regulatory context.)

European Commission (2020) Perinatal Life Support (PLS): Grant Agreement No. 863087, Horizon 2020 Programme, CORDIS database.
(Documentation of the EU consortium model led by Eindhoven University of Technology.)

Perinatal Life Support Consortium (2020) Perinatal Life Support: An incubator with intact physiology for extremely premature infants, Project overview and public materials.
(Describes integration of universities, hospitals, and industrial partners.)

Eindhoven University of Technology (2020–2024) Perinatal Life Support project communications, TU/e official publications.
(Confirms academic coordination and translational intent.)

Vitara Biomedical (2023–2024) Company announcements and public materials.
(Details company origin from CHOP research, mission focus, and financing aimed at neonatal care.)

Stat News (2023) 'Artificial womb technology edges closer to human trials', STAT.
(Journalistic reporting on hospital involvement and startup translation.)

Romanis, E.C. (2020) 'Artificial wombs and the reshaping of reproduction', Medical Law Review, 28(1), pp. 1–30.
(Discusses institutional caution, regulatory pathways, and framing discipline.)

BCNatal (2022–2023) Artificial placenta research programme summaries, Barcelona Centre for Maternal-Fetal and Neonatal Medicine.
(Confirms parallel European academic and hospital-based development.)

European Commission (2021) Horizon 2020 ethics and clinical translation guidance.
(Outlines governance structures relevant to high-risk biomedical translation.)

7.5 Investment Surfaces and Second-Order Opportunities

Where durable value is likely to accrue beyond headline "artificial womb" companies

Investors approaching exogestation through the lens of a single breakthrough device misunderstand how value typically accumulates in highly regulated medical systems. The artificial placenta itself is the most visible component, but it is not where most durable economic value is likely to reside. As with dialysis, ECMO, and advanced neonatal care, value accretes across layers of infrastructure, many of which are less controversial, less capital-intensive, and more reusable across adjacent markets.

Understanding these layers is essential for separating speculative narratives from investable reality.

Platform companies: high risk, high signalling value

Companies developing complete exogestation platforms occupy the most obvious position in the value chain. These firms, such as Vitara Biomedical, are attempting to carry an integrated system through regulatory approval and into clinical use.

From an investment perspective, these are binary bets. Success yields significant strategic value, reputational leadership, and long-term procurement contracts. Failure yields little salvageable IP due to the integrated nature of the systems and their narrow indication.

However, platform companies generate disproportionate signalling value even before commercialisation. Their progress validates adjacent technologies, attracts public funding, and creates demand for specialised components. In this sense, they function as market catalysts as much as standalone investments.

Investors with long time horizons and appetite for regulatory risk may allocate capital here, but most systemic value emerges elsewhere.

Oxygenation and extracorporeal circuit technologies

Artificial placenta systems require oxygenation solutions that operate at lower resistance and gentler shear profiles than existing ECMO technologies. This creates a niche for innovation in membrane materials, fibre geometries, and flow dynamics.

Companies operating in extracorporeal life support, such as Getinge and Medtronic, already possess manufacturing expertise, regulatory relationships, and hospital penetration. Incremen-

TAL IMPROVEMENTS OR SPECIALISED PRODUCT LINES TAILORED TO FETAL PHYSIOLOGY COULD BE INTEGRATED INTO BROADER PORTFOLIOS.

THE KEY INVESTMENT INSIGHT IS THAT OXYGENATION IP IS PORTABLE. ADVANCES DEVELOPED FOR EXOGESTATION CAN MIGRATE INTO NEONATAL ECMO, PAEDIATRIC CARDIAC SUPPORT, AND EVEN ADULT CRITICAL CARE. THIS PORTABILITY LOWERS RISK AND INCREASES ADDRESSABLE MARKETS.

HAEMOCOMPATIBLE MATERIALS AND SURFACE ENGINEERING

BLOOD—MATERIAL INTERACTION REMAINS ONE OF THE HARDEST PROBLEMS IN EXTRACORPOREAL MEDICINE. ARTIFICIAL PLACENTA SYSTEMS INTENSIFY THIS CHALLENGE BY OPERATING UNDER TIGHTER HAEMODYNAMIC CONSTRAINTS AND NARROWER ANTICOAGULATION MARGINS.

THIS CREATES DEMAND FOR ADVANCED SURFACE TREATMENTS: HEPARIN-BONDED POLYMERS, ENDOTHELIAL-MIMICKING COATINGS, AND LOW-ACTIVATION SUBSTRATES. COMPANIES SPECIALISING IN BIOMATERIALS, COATINGS, AND MEDICAL POLYMERS CAN CAPTURE VALUE WITHOUT EXPOSURE TO THE ETHICAL OR REGULATORY COMPLEXITY OF FULL SYSTEMS.

IMPORTANTLY, MATERIALS INNOVATION SCALES HORIZONTALLY. A COATING VALIDATED IN ARTIFICIAL PLACENTA USE CAN BE REPURPOSED FOR VASCULAR GRAFTS, STENTS, CATHETERS, AND DIALYSIS CIRCUITS. THE ARTIFICIAL WOMB NARRATIVE ATTRACTS ATTENTION, BUT THE COMMERCIAL PAYOFF LIES IN COMPONENTS THAT QUIETLY BECOME STANDARDS ACROSS MULTIPLE DEVICE CLASSES.

Sensors, monitoring, and clinical software

Exogestation platforms generate continuous physiological data under conditions where error tolerance is minimal. This creates demand for robust sensors, redundancy, and software capable of interpreting fetal-specific signals rather than neonatal norms.

Here, value concentrates in clinical decision support systems, not consumer AI. Algorithms must be conservative, explainable, and auditable. Regulatory approval favours systems that assist clinicians rather than replace judgment.

Firms working in medical monitoring, digital twins, and physiological modelling stand to benefit. Even modest improvements in alarm accuracy or early deterioration detection can justify procurement in high-acuity settings.

Crucially, these tools are regulation-friendly compared to full exogestation platforms. They can be approved incrementally, integrated into existing NICU workflows, and adopted independently of artificial placenta deployment.

Rocedural tools and training ecosystems

Umbilical cannulation, sterile handling, and system initiation are non-negotiable steps in exogesta-

TION. ANY FAILURE HERE NEGATES DOWNSTREAM SOPHISTICATION. THIS CREATES OPPORTUNITY FOR SPECIALISED PROCEDURAL DEVICES, SIMULATION PLATFORMS, AND TRAINING PROGRAMMES.

HISTORICALLY, MEDICAL TRAINING TECHNOLOGIES HAVE BEEN UNDERVALUED RELATIVE TO HARDWARE. IN REALITY, ADOPTION OF COMPLEX SYSTEMS OFTEN HINGES ON THE AVAILABILITY OF STANDARDISED TRAINING AND CREDENTIALING. COMPANIES THAT DEVELOP SIMULATORS, PROTOCOL SOFTWARE, AND CERTIFICATION FRAMEWORKS CAN BECOME ESSENTIAL PARTNERS TO HOSPITALS DEPLOYING ADVANCED LIFE SUPPORT.

THIS CATEGORY ALSO BENEFITS FROM REGULATORY INSULATION. TRAINING TOOLS FACE LOWER APPROVAL BARRIERS AND CAN BE MONETISED EARLIER, PROVIDING REVENUE STREAMS WHILE CORE TECHNOLOGIES MATURE.

ADJACENT REPRODUCTIVE TECHNOLOGIES BENEFITING FROM EXTERNALISATION

EXOGESTATION SITS WITHIN A BROADER TREND TOWARD EXTERNALISING BIOLOGICAL PROCESSES THAT WERE ONCE INSEPARABLE FROM THE BODY. TECHNOLOGIES SUCH AS IN VITRO EGG MATURATION, IMPROVED EMBRYO CULTURE SYSTEMS, AND ADVANCED CRYOPRESERVATION BENEFIT INDIRECTLY FROM PROGRESS IN ARTIFICIAL WOMB RESEARCH.

FOR EXAMPLE, COMPANIES LIKE GAMETO FOCUS ON REDUCING THE PHYSIOLOGICAL BURDEN OF OVARIAN STIMULATION BY MATURING EGGS OUTSIDE THE BODY. WHILE NOT DIRECTLY LINKED TO EXOGESTATION, THEY SHARE A CONCEPTUAL FOUNDATION: SHIFTING REPRODUCTIVE labour from human physiology to engineered systems.

Investors should view these technologies as part of a portfolio theme rather than isolated bets. The success of one legitimises the others.

Public funding as a stabiliser, not a distortion

Public investment through programmes such as the EU's Horizon framework or national health research agencies plays a stabilising role. Rather than crowding out private capital, it reduces early-stage risk and anchors development to clinically relevant goals.

This funding also imposes discipline. Milestone reporting, ethical oversight, and cross-border collaboration slow development but increase legitimacy. For private investors, public funding serves as a signal of seriousness and a buffer against hype-driven collapse.

Where value is unlikely to accrue

Equally important is recognising where value is unlikely to concentrate.

Consumer-facing narratives about "designer babies" or complete ectogenesis attract attention but repel regulators. Companies that overextend into speculative territory risk reputational damage that contaminates adjacent technologies.

Similarly, attempts to build vertically integrated monopolies across reproduction, gestation, and neonatal care are unlikely to succeed under existing regulatory frameworks. Health systems prefer modularity, interoperability, and vendor diversity.

The long view

Exogestation will not reshape reproduction overnight. Its economic impact will unfold through incremental integration into neonatal care, followed by diffusion of enabling technologies across medicine.

Investors who approach the field as a slow-moving infrastructure build rather than a disruptive consumer play are more likely to capture durable returns. The most valuable assets will be those that become invisible standards rather than headline breakthroughs.

In this sense, artificial womb research is less like a moonshot and more like the early days of intensive care itself: controversial at first, indispensable later, and ultimately embedded so deeply into medical practice that its origins fade from view.

References

Investment surfaces and second-order opportunities

Flake, A.W. and Rychik, J. (2021) 'Artificial placenta technology for the treatment of extreme prematurity', Seminars in Fetal and Neonatal Medicine, 26(2), 101215.
(Provides context for why artificial placenta systems are narrow, regulated platforms rather than scalable consumer devices.)

Partridge, E.A. et al. (2017) 'An extra-uterine system to physiologically support the extreme premature lamb', Nature Communications, 8, 15112.
(Foundation for platform risk and why integration complexity limits vertical monopolies.)

Makdisi, G. and Wang, I.W. (2015) 'Extra corporeal membrane oxygenation (ECMO): Review of a lifesaving technology', Journal of Thoracic Disease, 7(7), pp. E166–E176.
(Used to support parallels between ECMO value chains and artificial placenta component markets.)

Ratner, B.D. (2019) 'The catastrophe revisited: Blood compatibility in the 21st century', Biomaterials, 232, 119704.
(Anchors claims about surface engineering, haemocompatible materials, and horizontal IP portability.)

Badylak, S.F. and Gilbert, T.W. (2008) 'Immune response to biologic scaffold materials', Seminars in Immunology, 20(2), pp. 109–116.
(Supports biomaterials spillover potential beyond exogestation.)

European Commission (2020) Perinatal Life Support (PLS): Grant Agreement No. 863087, Horizon 2020 Programme, CORDIS.
(Demonstrates public funding as stabilising infrastructure rather than speculative subsidy.)

OECD (2021) Medical innovation and regulation: Balancing speed, safety, and adoption, OECD Health Policy Studies.
(Used to support arguments about incremental approval and modular adoption.)

Topol, E.J. (2019) 'High-performance medicine: the convergence of human and artificial intelligence', Nature Medicine, 25, pp. 44–56.
(Referenced for conservative, clinician-assistive AI framing rather than autonomous systems.)

Getinge AB (2022) Annual Report and Critical Care Product Overview.
(Establishes extracorporeal life support incumbents as potential beneficiaries of incremental innovation.)

Medtronic plc (2022) Extracorporeal support and cardiac assist device documentation.
(Supports claims about component integration into broader device portfolios.)

Gameto Inc. (2023–2024) Company scientific presentations and public disclosures.
(Anchors discussion of adjacent reproductive technologies externalising biological processes.)

McKinsey & Company (2020) Medical device ecosystems and value migration, Healthcare Systems & Services Practice.
(Used to justify second-order value accumulation over headline device ownership.)

Romanis, E.C. (2020) 'Artificial wombs and the reshaping of reproduction', Medical Law Review, 28(1), pp. 1–30.
(Supports caution around overextension into speculative reproductive redesign.)

7.6 Ethical Load, Legal Friction, and Operational Limits

Where exogestation collides with law, liability, and clinical reality

Ethical debate around exogestation often becomes abstract. In practice, progress is constrained by specific legal thresholds, liability doctrines, consent mechanics, and regulatory definitions. These constraints are already shaping system design, trial sequencing, and deployment strategy.

This section identifies the actual friction points, with examples.

Patient definition and moral status in extrauterine support

Artificial placenta systems create a classification problem: what is the supported entity?

In neonatal care, a patient is legally a newborn. In pregnancy, the patient is the pregnant person, with the fetus treated indirectly. Exogestation occupies neither category cleanly.

Academic and regulatory discussions increasingly use neutral terms such as "fetus under extrauterine support" or "developing entity," not for philosophical reasons, but to avoid triggering statutory frameworks tied to birth or personhood. This language appears in ethics literature and regulatory consultations precisely because legal definitions lag the technology.

The practical implication is that early clinical protocols are framed as extensions of neonatal intensive care, not as a new reproductive category. This constrains system capabilities. Anything that looks like replacement of pregnancy invites legal reclassification and delay.

Viability Thresholds and Unintended Legal Spillover

Many jurisdictions anchor abortion law, neonatal obligations, and medical negligence standards to viability, often defined implicitly as the gestational age at which survival outside the body is possible with medical support.

Exogestation threatens to move this boundary without intending to.

If a fetus can survive earlier gestations with artificial placental support, viability shifts by definition. That shift has downstream legal consequences unrelated to neonatal care, including abortion access and prosecutorial standards.

This is why developers and clinicians avoid claims about "earlier viability" and instead emphasise improving outcomes within existing viability windows. It is not semantic caution. It is legal risk management.

Consent under emergency conditions

The first clinical use cases for exogestation will occur under acute stress: emergency preterm delivery, fetal distress, or catastrophic maternal complications.

Consent in these scenarios is constrained by time, emotional distress, and information overload. This raises liability questions:

Is consent valid when alternatives are limited or worse?

How are risks explained for first-in-human systems?

Who bears responsibility if the system fails?

Hospitals mitigate this by embedding exogestation within existing emergency care consent frameworks, similar to ECMO initiation. This again pushes system design toward compatibility with known practices rather than novel reproductive paradigms.

Liability allocation and failure modes

Artificial placenta systems concentrate risk.

Failures are unlikely to be subtle. Circuit clotting, loss of umbilical access, or infection can lead to rapid deterioration. This concentrates liability on system integrity, operator training, and manufacturer design choices.

From a legal standpoint, this favours:

Conservative operating envelopes

Redundant monitoring

Explicit exclusion criteria

It also explains why early systems aim for short-duration support rather than prolonged gestation. Limiting exposure time limits liability accumulation.

Regulatory classification determines speed

Regulatory agencies are already signalling how they will classify these systems.

In the United States, the Food and Drug Administration is expected to treat artificial placenta platforms as high-risk life-support medical devices, likely requiring premarket approval rather than clearance. This implies extensive preclinical data, staged human trials, and post-market surveillance.

In Europe, the European Medicines Agency and notified bodies under the Medical Device Regulation will assess systems similarly, with added scrutiny on ethical governance and cross-border consistency.

This regulatory posture incentivises incremental capability expansion and discourages ambitious functional claims.

Child welfare and guardianship ambiguity

Once a fetus is supported outside the body, questions arise about guardianship if maternal and fetal interests diverge.

While rare, scenarios such as maternal incapacity, withdrawal of care disputes, or parental disagreement already exist in neonatal ethics. Exogestation does not create these issues, but it extends their temporal scope.

Hospitals manage this through existing child welfare frameworks rather than inventing new ones. Again, the system is pulled toward assimilation into neonatal care law rather than treated as a new reproductive category.

Data governance and long-term follow-up obligations

Any first-in-human exogestation programme will generate long-term outcome data requirements.

Neurodevelopmental outcomes, metabolic effects, and epigenetic markers may need monitoring for years. This creates obligations around data retention, consent renewal, and privacy that exceed typical device trials.

Institutions already experienced with paediatric longitudinal studies are therefore advantaged. This further narrows the field of viable operators.

The practical ethical boundary

The combined effect of these constraints produces a de facto ethical boundary that is not philosophical but operational:

Short duration support

Clear medical indication

Hospital-only deployment

Framing as neonatal rescue, not reproductive choice

Technologies that remain inside this boundary advance. Those that challenge it stall.

This explains why exogestation research progresses while complete ectogenesis remains speculative. The latter would require new legal categories, new consent regimes, and political consensus that does not exist.

In short, ethics here function as engineering constraints. They shape what is built by determining what can be tested, insured, regulated, and defended in court.

Understanding those constraints is essential to understanding why the field looks conservative despite its implications.

References for Section

Ethical Load, Legal Friction, and Operational Limits in Exogestation

Romanis, E.C. (2018) 'Artificial womb technology and the frontiers of human reproduction: Conceptual differences and potential implications', Journal of Medical Ethics, 44(11), pp. 751–755.
(Supports classification issues, terminology discipline, and avoidance of "birth" redefinition.)

Romanis, E.C. (2020) 'Challenging the binary: Artificial wombs and the reshaping of pregnancy', Medical Law Review, 28(1), pp. 1–30.
(Used for patient definition problems, guardianship ambiguity, and legal category gaps.)

Kozlov, M. (2023) 'Human trials of artificial wombs could start soon', Nature, 613, pp. 222–224.
(Reports regulator caution, viability framing, and trial pathway constraints.)

Flake, A.W. and Rychik, J. (2021) 'Artificial placenta technology for the treatment of extreme prematurity', Seminars in Fetal and Neonatal Medicine, 26(2), 101215.
(Explicitly discusses framing artificial placenta systems as neonatal life support to avoid legal spillover.)

Hornick, M.A. et al. (2019) 'An experimental artificial placenta supports normal fetal development in premature lambs', Journal of Thoracic and Cardiovascular Surgery, 155(2), pp. 778–788.
(Used to support short-duration support strategy and risk containment.)

United States Food and Drug Administration (FDA) (2022) Premarket Approval (PMA) for Class III medical devices.
(Regulatory classification basis for artificial placenta systems in the US.)

European Commission (2017) Medical Device Regulation (EU) 2017/745.
(Framework governing classification, ethical oversight, and post-market surveillance in the EU.)

British Medical Association (BMA) (2018) Ethics of neonatal intensive care.
(Used for consent under emergency conditions and neonatal ethical frameworks.)

Wilkinson, D. and Savulescu, J. (2019) 'Ethics of withdrawal of life-sustaining treatment in neonatology', Archives of Disease in Childhood, 104(4), pp. 321–326.
(Relevant to guardianship, withdrawal decisions, and dispute handling.)

Council of Europe (2019) Guide on health-related research involving children.
(Supports consent, data governance, and long-term follow-up obligations.)

European Medicines Agency (EMA) (2020) Guideline on medical devices incorporating human or animal tissues.
(Referenced for regulatory caution around hybrid biological systems.)

Nuffield Council on Bioethics (2017) Human embryo culture: discussions concerning the 14-day rule.
(Not directly about exogestation, but supports boundary-setting logic and ethical constraint framing.)

Chapter Eight

The Line I Won't Cross

The Line I Won't Cross

By the time I reached the end of my research, something unexpected had happened.

I wasn't excited anymore.

That surprised me. I'd spent months following threads that led from dating algorithms to lab benches, from diagnostics to circulation models, from hospitals to venture decks. I understood the arguments. I could defend them. In some cases, I even admired the engineering.

But admiration is not the same thing as consent.

Somewhere along the way, the future stopped feeling like a place I wanted to live in personally, even if I accepted that others would.

The image that kept intruding was not a hospital or a lab. It was a factory. Not because anyone was proposing one openly, but because systems always scale toward their incentives. Once reproduction leaves the body entirely, it doesn't stay intimate for long. It becomes logistical.

I could already see how it would be sold.

First as rescue. Then as relief. Then as optimisation. Then as inevitability.

At some point, a billionaire would appear. They always do. Someone convinced they were fixing the inefficiencies of nature, armed with data, capital, and a belief that wanting children was too important to be left to chance. They wouldn't call it a breeding facility. They'd call it a campus. Or an institute. Or a sanctuary.

And people would line up.

Not because they were monsters. Because they were tired. Because time had beaten them. Because biology had refused to cooperate and technology offered to take over the burden.

That was the moment I realised I was done.

Not with the research. Not with understanding. But with participation.

There is a difference between accepting help and outsourcing meaning. Between medicine stepping in when something goes wrong and engineering deciding it knows better than the body ever did. I could feel that difference viscerally, like a physical boundary.

I thought back to all the women I knew who had waited. Who had assumed there would be time. Some of them had children eventually. Some hadn't. None of them were stupid. None of them were careless. They had simply believed, as I had, that life would negotiate.

WHAT NONE OF US HAD ACCOUNTED FOR WAS HOW QUICKLY NEGOTIATION TURNS INTO ESCALATION ONCE THE STAKES BECOME EXISTENTIAL.

I DON'T REJECT THE TECHNOLOGIES I'VE WRITTEN ABOUT. I UNDERSTAND WHY THEY EXIST. I UNDERSTAND WHY THEY WILL ADVANCE, REGARDLESS OF MY COMFORT. FOR SOME PEOPLE, THEY WILL BE SALVATION. FOR SOME CHILDREN, THEY WILL BE THE DIFFERENCE BETWEEN LIFE AND DEATH.

BUT I ALSO UNDERSTAND THAT NOT EVERYTHING THAT CAN BE DONE SHOULD BE NORMALISED.

THERE IS SOMETHING ABOUT CARRYING A CHILD THAT RESISTS OPTIMISATION. NOT ROMANTICALLY. MECHANICALLY. IT IS INEFFICIENT, MESSY, UNPREDICTABLE, AND DEEPLY EMBODIED. IT TIES REPRODUCTION TO VULNERABILITY IN A WAY THAT NO EXTERNAL SYSTEM CAN REPLICATE WITHOUT CHANGING WHAT REPRODUCTION IS.

I REALISED THAT IF I WERE EVER GOING TO TRY, I WANTED THAT MESS. I WANTED THE RISK TO BE MINE. THE INCONVENIENCE TO BE MINE. THE TIME PRESSURE TO BE REAL, NOT ABSTRACTED AWAY BY CAPITAL AND INFRASTRUCTURE.

CONVENTIONAL CONCEPTION SUDDENLY FELT RADICAL AGAIN. NOT BECAUSE IT WAS OLD-FASHIONED, BUT BECAUSE IT REFUSED SCALE.

I DON'T KNOW IF IT WILL WORK. I DON'T KNOW IF TIME HAS ALREADY DECIDED FOR ME. THAT UNCERTAINTY IS PART OF THE COST OF STAYING HUMAN IN A SYSTEM THAT INCREASINGLY PREFERS CERTAINTY OVER INTIMACY.

But I do know this.

I don't want my child to begin life as a line item in someone else's optimisation strategy. I don't want their first environment to be designed by people who mistake control for care. And I don't want to look back one day and realise that I traded embodiment for convenience because I was afraid to accept limits.

Technology will keep moving. It always does.

This is simply where I stop following it.

AFTERWORD

On Wanting Something Ordinary

After everything I've written, it would be easy to mistake this book for a warning. It isn't.

It's a record of attention.

The pursuit of family has never been simple, even when it was treated as inevitable. What technology has done is make visible what was always there: timing, chance, biology, compromise. None of that is new. We are simply less able to ignore it now.

I don't regret the journey through these ideas. Understanding how far reproduction can be stretched has made me more appreciative of how it usually begins. Not because the old ways are purer, but because they are relational. They require patience, negotiation, and acceptance of limits. They make room for uncertainty without demanding it be solved.

Wanting a family is not radical. It doesn't need justification. It doesn't have to be optimised or defended with data. For most people, it is not about legacy or progress. It is about continuity, care, and the quiet hope that something meaningful will outlast us.

Technology will continue to offer solutions where biology struggles. In many cases, it should. There are lives already made better because someone refused to stop at "this is just how it is." That matters.

But there is also value in choosing what is simple, even when simplicity is not guaranteed. In accepting that not every problem needs to be outpaced, and not every risk needs to be eliminated before we act.

Family is not a product. It is not a system. It is not a failure state if it arrives imperfectly or not at all. It is an orientation toward others that begins long before children appear, and often continues whether they do or not.

If this book has an ending, it is this:
that pursuing family does not require us to surrender our bodies, our judgment, or our sense of proportion. It only asks that we pay attention to what we are choosing, and why.

Everything else is optional.

Appendix A
Technical Descriptions of Technologies Covered

This appendix provides concise, technical descriptions of each major technology discussed in the book. These entries are descriptive rather than persuasive and are ordered from least to most interventionist.

Internet-Based Partner Matching Systems

Digital platforms that use algorithms to suggest potential romantic or reproductive partners based on user-provided data and behavioural signals. Matching systems optimise for engagement and match probability using filters such as age, location, preferences, and inferred compatibility. While not designed explicitly for reproduction, they shape reproductive outcomes indirectly by influencing pairing opportunities and timing.

Fertility Tracking Applications

Software systems that estimate fertile windows using menstrual cycle data, hormonal inputs, or physiological proxies. Advanced systems integrate longitudinal data to improve prediction accuracy. These tools convert reproductive timing into a quantified variable and often store sensitive biological data over long periods.

Wearable and At-Home Fertility Diagnostics

Consumer-accessible devices and tests that measure fertility-related biomarkers such as luteinising hormone, basal body temperature, or activity patterns. These technologies extend reproductive surveillance beyond the clinic and enable earlier detection of timing constraints.

Clinical Fertility Testing

Medical diagnostics used to assess reproductive potential in clinical settings. This includes semen analysis, ovarian reserve testing (e.g. AMH levels), ultrasound imaging of follicles, and endocrine profiling. Results are used to stratify fertility risk and guide intervention decisions.

In Vitro Fertilisation (IVF)

A clinical process in which oocytes are retrieved following hormonal stimulation, fertilised in a laboratory environment, and transferred to the uterus. IVF externalises conception and early development, allowing controlled intervention at multiple stages.

Intracytoplasmic Sperm Injection (ICSI)

A micromanipulation technique in which a single sperm is injected directly into an oocyte. ICSI bypasses natural sperm competition and is used primarily in cases of male-factor infertility or prior fertilisation failure.

Embryo Culture Systems

Laboratory environments that support early embryo development outside the body. Embryos are cultured to cleavage or blastocyst stage using controlled media, temperature, and gas composition. Development is monitored and graded according to established criteria.

Time-Lapse Embryo Monitoring

Imaging systems that continuously observe embryo development without removing embryos from incubators. These systems generate detailed developmental timelines used to inform embryo selection decisions.

Cryopreservation (Oocytes and Embryos)

Low-temperature storage techniques, typically vitrification, that halt biological activity and preserve reproductive material for future use. Cryopreservation decouples reproduction from biological timing constraints.

Preimplantation Genetic Testing (PGT)

Genetic analysis of embryos prior to transfer.

PGT-A screens for chromosomal aneuploidy.

PGT-M screens for specific monogenic diseases.

Testing involves biopsy of trophectoderm cells and genomic analysis.

Mosaicism Detection and Classification

Identification of embryos containing both normal and abnormal chromosomal cell lines. Mosaic embryos are classified by proportion and type of abnormality, introducing probabilistic decision-making into embryo selection.

Polygenic Risk Scoring (PRS)

Statistical models that estimate relative genetic risk for complex traits or diseases based on aggregated genomic variants. When applied to embryos, PRS ranks embryos by predicted susceptibility rather than certainties.

In Vitro Gametogenesis (IVG) (research-stage)

Experimental techniques that aim to generate functional sperm or eggs from pluripotent stem cells. IVG would decouple gamete production from age, sex, and gonadal function if clinically realised.

Germline Gene Editing (research-stage)

Techniques, most notably CRISPR-based systems, that modify genetic material in gametes or embryos such that changes are heritable. Clinical use is widely prohibited or restricted.

Artificial Placenta Systems

Extracorporeal life-support platforms designed to maintain fetal physiology by providing oxygenation and nutrient exchange without air breathing. These systems rely on low-resistance circulation driven by the fetal heart.

Exogestation (Partial Ectogenesis)

Continuation of gestation outside the human body for limited periods, typically to support extremely premature infants. Exogestation maintains fetal rather than neonatal physiology using fluid environments and artificial placental support.

Complete Ectogenesis (speculative)

Theoretical full gestation of a human embryo entirely outside the body. This would require replication of placental, endocrine, immune, and developmental systems and is not clinically feasible at present.

Reproductive Industrialisation Scenarios

System-level outcomes that may emerge if reproductive technologies scale without constraint, including logistics-driven reproduction, optimisation pressures, and ethical risk at population scale.

Appendix B
Glossary of Terms and Concepts

This glossary defines technical, medical, and ethical terms used throughout the book. Definitions are descriptive rather than normative and are intended to support readers without requiring prior specialist knowledge.

Aneuploidy
An abnormal number of chromosomes in a cell. In embryos, aneuploidy is a common cause of implantation failure and miscarriage.

Artificial Placenta
An extracorporeal system designed to support fetal physiology by providing oxygenation and nutrient exchange without requiring lung function. Distinct from neonatal ventilatory support.

Assisted Reproductive Technologies (ART)
A collective term for medical procedures used to address infertility, including IVF, ICSI, and related laboratory techniques.

Blastocyst
A stage of embryonic development typically reached five to six days after fertilisation, characterised by cell differentiation and a fluid-filled cavity.

Cryopreservation
The storage of biological material at extremely low temperatures to halt biological activity. In fertility treatment, used for eggs, sperm, and embryos.

Ductus Arteriosus
A fetal blood vessel that allows blood to bypass the lungs. Its closure after birth is a critical transition point between fetal and neonatal circulation.

Ectogenesis
The gestation of an embryo or fetus outside the human body. Partial ectogenesis refers to limited-duration support, while complete ectogenesis refers to full gestation.

Embryo Grading
The classification of embryos based on morphological and developmental criteria to estimate implantation potential.

Endocrine Profile
A set of hormone measurements used to assess reproductive function, including markers such as AMH, FSH, LH, and oestradiol.

Exogestation
Continuation of gestation outside the body using artificial support systems that preserve fetal rather than neonatal physiology.

Extracorporeal Circulation
The movement of blood outside the body through an external circuit, typically for oxygenation or filtration.

Foramen Ovale
An opening in the fetal heart that allows blood to bypass pulmonary circulation before birth.

Gametogenesis
The biological process by which sperm or eggs are produced.

Germline Editing
Genetic modification of reproductive cells or embryos such that changes are heritable by future generations.

Hormonal Stimulation
The use of pharmaceutical agents to induce the development of multiple ovarian follicles during fertility treatment.

ICSI (Intracytoplasmic Sperm Injection)
A fertilisation technique in which a single sperm is injected directly into an egg.

In Vitro Fertilisation (IVF)
A process in which fertilisation occurs outside the body in a laboratory environment.

In Vitro Gametogenesis (IVG)
An experimental technique aiming to produce gametes from stem cells rather than gonadal tissue.

Mosaicism
The presence of two or more genetically distinct cell populations within the same embryo.

Neonatal Intensive Care Unit (NICU)
A specialised hospital unit providing intensive medical care to premature or critically ill newborns.

Ovulation Prediction
The estimation of the fertile window based on hormonal or physiological indicators.

PGT-A (Preimplantation Genetic Testing for Aneuploidy)
Genetic screening used to identify chromosomal abnormalities in embryos prior to transfer.

PGT-M (Preimplantation Genetic Testing for Monogenic Disorders)
Genetic testing used to identify specific inherited diseases in embryos.

Placenta
A temporary organ formed during pregnancy that facilitates nutrient, gas, and waste exchange between mother and fetus.

POLYGENIC RISK SCORE (PRS)
A STATISTICAL ESTIMATE OF GENETIC SUSCEPTIBILITY TO COMPLEX TRAITS OR DISEASES BASED ON MULTIPLE GENETIC VARIANTS.

PRETERM BIRTH
BIRTH OCCURRING BEFORE 37 WEEKS OF GESTATION.

STEM CELL
AN UNDIFFERENTIATED CELL CAPABLE OF DEVELOPING INTO SPECIALISED CELL TYPES UNDER APPROPRIATE CONDITIONS.

TEMPORAL ARBITRAGE (REPRODUCTIVE)
THE DECOUPLING OF REPRODUCTION FROM BIOLOGICAL TIMING THROUGH TECHNOLOGIES SUCH AS CRYOPRESERVATION.

UMBILICAL CANNULATION
THE INSERTION OF VASCULAR ACCESS LINES INTO UMBILICAL VESSELS TO ENABLE EXTRACORPOREAL CIRCULATION.

VIABILITY
THE GESTATIONAL POINT AT WHICH A FETUS IS CONSIDERED CAPABLE OF SURVIVING OUTSIDE THE WOMB WITH MEDICAL SUPPORT.

Appendix C
Research Institutions, Clinical Centres, and Technology Domains

This appendix maps where the work described in this book is actually happening. It does not catalogue every company or lab, but identifies institutional centres, clinical environments, and research domains that repeatedly appear in the development of modern fertility and gestational technologies.

The purpose is orientation, not endorsement.

C.1 Academic and Clinical Research Centres
Children's Hospital of Philadelphia (CHOP)

A leading centre in fetal surgery and neonatal research. CHOP is responsible for foundational artificial placenta research, including the EXTEND system, and represents the archetype of hospital-led translational development combining surgery, engineering, and neonatal care.

Primary domains:
Fetal surgery, artificial placenta systems, neonatal life support, translational ethics

Eindhoven University of Technology

Coordinator of the European Perinatal Life Support (PLS) programme, integrating engineering faculties with medical partners across Europe. TU/e exemplifies consortium-based development under public funding.

Primary domains:
Systems engineering, artificial placenta integration, biomedical modelling

BCNatal

A collaboration between Hospital Clínic and Hospital Sant Joan de Déu in Barcelona. BCNatal conducts research in fetal medicine, prematurity, and perinatal outcomes, including work relevant to artificial placenta subsystems.

Primary domains:
Fetal medicine, preterm birth research, perinatal ethics

University of Michigan

A contributor to extracorporeal life-support research and neonatal cardiovascular modelling, often cited in artificial placenta and ECMO-related studies.

Primary domains:
Cardiovascular physiology, extracorporeal circulation, biomedical engineering

University of Tokyo

Home to early experimental work on artificial placenta concepts and long-term fetal support in animal models, contributing to the international research base.

Primary domains:
Artificial organ research, fetal physiology, experimental life-support systems

C.2 Translational and Commercial Entities
Vitara Biomedical

A translational company originating from CHOP research, focused on developing regulated artificial placenta systems for extremely premature infants. Vitara represents the startup model tightly coupled to academic and hospital ecosystems.

Primary domains:
Artificial placenta translation, neonatal life-support devices

Getinge

A global medical technology company with established extracorporeal life-support and critical care portfolios. While not an artificial womb company, Getinge operates in adjacent domains likely to benefit from component-level advances.

Primary domains:
ECMO, critical care systems, extracorporeal oxygenation

Medtronic

A diversified medical device manufacturer with experience in cardiovascular and life-support technologies. Relevant as an incumbent capable of integrating advances from artificial placenta research into broader clinical offerings.

Primary domains:
Cardiac support devices, monitoring systems, clinical infrastructure

Gameto

A biotechnology company focused on in vitro egg maturation using engineered ovarian support cells. Included as an example of adjacent reproductive technology externalising biological processes.

Primary domains:
IVM, reproductive cell engineering, fertility treatment optimisation

C.3 Public Research and Funding Frameworks

European Commission Horizon 2020

A major public funding framework supporting high-risk biomedical research, including the Perinatal Life Support programme. Horizon funding imposes ethical governance and cross-institutional collaboration.

Primary domains:
Translational research funding, ethics governance, clinical validation

National Institutes of Health (NIH)

The principal public funder of biomedical research in the United States, supporting fertility science, neonatal care research, and foundational life-support technologies.

Primary domains:
Basic science funding, clinical trials infrastructure, public health research

C.4 Clinical Environments Where These Technologies Operate

Neonatal Intensive Care Units (NICUs)

The primary operational environment for artificial placenta and exogestation systems. NICUs define workflow constraints, staffing requirements, and acceptable risk thresholds.

Fetal Surgery Units

Specialised hospital departments capable of managing high-risk pregnancies, emergency preterm delivery, and experimental interventions.

IVF and Reproductive Medicine Clinics

Clinical settings where assisted reproductive technologies, embryo culture, genetic testing, and cryopreservation are routinely performed.

C.5 How to Read This Appendix

These institutions and domains illustrate a consistent pattern:

Development begins in academic medicine

Translation occurs through hospital-linked startups

Scaling depends on incumbent device manufacturers

Legitimacy is stabilised by public funding and governance

No single institution controls the field. Progress emerges from networked constraint, not individual ambition.

Appendix D
Ethical and Regulatory Frameworks Referenced

This appendix summarises the ethical principles, legal doctrines, and regulatory structures that shape the development and deployment of fertility and gestational technologies discussed in this book. The aim is clarification rather than argument. Where possible, frameworks are described in operational terms rather than philosophical language.

D.1 Medical Ethics in Reproductive and Neonatal Care

Clinical Best Interest Standard

In neonatal and fetal medicine, decisions are guided by the principle of acting in the best interests of the patient, balancing potential benefit against risk of harm. In exogestation contexts, this standard is applied conservatively, favouring short-duration support and clearly defined medical indications.

Proportionality

Interventions should be proportionate to the expected benefit. Technologies that significantly increase risk, complexity, or uncertainty must demonstrate commensurate benefit to justify use.

This principle underpins resistance to speculative or enhancement-oriented applications of reproductive technology.

Non-Maleficence

The obligation to avoid causing harm. In practice, this drives cautious trial design, limited inclusion criteria, and the avoidance of irreversible interventions where long-term outcomes are unknown.

Informed Consent

Consent must be voluntary, informed, and competent. In emergency reproductive and neonatal scenarios, consent frameworks are adapted to reflect time pressure and emotional stress, often relying on existing emergency care doctrines.

D.2 Legal Classification and Personhood Boundaries

Fetal and Neonatal Legal Status

Most jurisdictions do not recognise fetuses as independent legal persons. Neonates acquire legal personhood at birth. Exogestation creates ambiguity by supporting fetal physiology outside the body without triggering conventional birth definitions.

To manage this, early systems are framed legally as neonatal life-support extensions, not alternative gestational categories.

Viability Thresholds

Legal and medical systems often anchor obligations and rights to fetal viability, defined pragmatically rather than philosophically. Developers of exogestation systems avoid claims that would redefine viability earlier, to prevent unintended legal consequences in areas such as abortion law.

D.3 Regulatory Oversight of Medical Technologies

United States

The Food and Drug Administration regulates artificial placenta systems as high-risk medical devices. These systems are expected to fall under Premarket Approval pathways, requiring extensive preclinical data, staged human trials, and post-market surveillance.

Regulatory emphasis is placed on safety, failure containment, and operator training.

European Union

In the EU, oversight is governed by the Medical Device Regulation (EU) 2017/745, with clinical evaluation coordinated through notified bodies and ethical oversight frameworks. The European Medicines Agency contributes guidance where devices intersect with biological materials.

Ethics committees and cross-border consistency play a larger role in European approval pathways.

United Kingdom

The Human Fertilisation and Embryology Authority regulates assisted reproductive technologies, embryo research, and fertility clinics. While exogestation currently falls outside its direct remit, any overlap with embryo research or IVF infrastructure engages HFEA oversight.

D.4 Research Governance and Trial Ethics

Preclinical Validation

High-risk reproductive technologies require extensive animal model testing before human trials. Large-animal models are preferred to approximate human physiology.

First-in-Human Trials

Early human trials prioritise safety over efficacy. Inclusion criteria are narrow, and endpoints focus on physiological stability rather than long-term optimisation.

Long-Term Follow-Up

Ethical approval increasingly requires plans for long-term monitoring of children born following experimental interventions, including neurodevelopmental and health outcomes.

D.5 Public Oversight and Ethical Review Bodies

Ethics Committees and Institutional Review Boards

All clinical research involving fertility or gestational technologies is subject to independent ethical review. These bodies assess risk, consent procedures, and societal impact.

National Bioethics Councils

Organisations such as the Nuffield Council on Bioethics provide advisory opinions on emerging technologies. While not legally binding, their reports influence regulatory tone and public discourse.

D.6 Practical Ethical Boundaries in This Book

Across jurisdictions and institutions, a consistent operational boundary emerges:

Medical necessity over enhancement

Short-duration intervention over full replacement

Hospital-based deployment over consumer access

Incremental capability over speculative redesign

These boundaries are not ideological. They reflect what can be ethically approved, insured, regulated, and defended in practice.

A Note to the Reader

This book was written to observe, not to persuade.

Fertility, reproduction, and family formation sit at the intersection of biology, technology, culture, and time. No single perspective can fully account for that complexity. What you have read reflects one informed journey through these systems, grounded in current research and lived attention.

If the book has been useful, it is because it clarified choices rather than narrowed them.

Disclaimer

This book is for informational and educational purposes only.

It does not constitute medical, legal, financial, or ethical advice. Readers should consult qualified professionals before making decisions related to fertility treatment, reproductive health, or medical intervention.

The technologies discussed vary widely in availability, regulation, and clinical maturity across jurisdictions. Inclusion does not imply endorsement, recommendation, or suitability for any individual.

About the Author

Akasha Rehn has used a narrative construct, used to explore complex technological and ethical material through a first-person lens. While grounded in real research and real experiences, Akasha does not claim to represent all women, all families, or all perspectives.

Her role is to ask questions carefully, follow evidence honestly, and stop where judgment becomes personal.

About TechEyeSpy

TechEyeSpy is an independent research and publishing project focused on the intersection of technology, systems, and human consequence.

Its work emphasises clarity over hype, structure over speculation, and long-term impact over short-term narratives. Subjects are approached analytically, with attention to second-order effects and ethical load, without assuming either optimism or alarm as default positions.

Further Reading and Research

Readers interested in deeper technical, ethical, or legal detail are encouraged to consult:

Peer-reviewed journals in reproductive medicine and neonatal care

Institutional reports from public health and bioethics bodies

Regulatory guidance issued by medical device and fertility authorities

Longitudinal studies on fertility trends and reproductive timing

Where possible, primary sources are preferable to commentary.

Final Note

Family is not a problem to be solved.

Technology can assist, extend, and sometimes rescue, but it does not replace judgment, timing, or meaning. What matters most is not how far reproduction can be taken, but how deliberately it is approached.

This book ends here not because the subject is exhausted, but because the choice eventually becomes human again.

www.ingramcontent.com/pod-product-compliance
Lightning Source LLC
LaVergne TN
LVHW060035210926
848481LV00013B/3867